MAPPING THE MIND

MAPPING THE MIND

THE INTERSECTION OF PSYCHOANALYSIS AND NEUROSCIENCE

Fred M. Levin

with a foreword by
John E. Gedo

Routledge
Taylor & Francis Group
LONDON AND NEW YORK

Earlier versions of the following chapters were published in *The Annual of Psychoanalysis* and appear here with the permission of the copyright holder, the Chicago Institute for Psychoanalysis: ch. 1 -Metaphor, Affect, and Arousal: How Interpretations Might Work (1980), 8:231-248; ch. 2 -Psychoanalysis and the Two Cerebral Hemispheres (1983) 11:171-197; ch. 3 -Brain Plast city, Learning, and Psychoanalysis (1987), 15:49-96.

An earlier version of chapter 6 appeared as the Introduction to *Repetition and Trauma,* by Max Stern (1988, The Analytic Press, Inc.).

Figures 1 and 4 in chapter 4 are reproduced with permission of M. Itoh;
Figure 2, with permission of McGraw-Hill Inc;
Figure 3, with permission of S. Niwa.

First published 1991 by The Analytic Press, Inc., Hillsdale, NJ

Published 2003 by Karnac Books Ltd.

2 Park Square, Milton Park, Abingdon, Oxfordshire OX14 4RN
52 Vanderbilt Avenue, New York, NY 10017

Routledge is an imprint of the Taylor & Francis Group, an informa business

First issued in paperback 2019

British Library Cataloguing in Publication Data
A C.I.P. for this book is available from the British Library

ISBN: 978-1-85575-300-6 (pbk)
ISBN: 978-0-367-32551-0 (hbk)

This book is dedicated to all those who have encouraged this kind of interdisciplinary research, especially the editors of *The Annual of Psychoanalysis,* who found merit in my endeavors long before anyone else did. I hope my productivity has justified their confidence in me. I am also deeply grateful to my wife, Sachiko, to my sons, David and Daniel, for their perfect blend of criticism and encouragement, and to my father, Harry Levin.

Acknowledgments

Mapping the Mind is written as a guide to those who care about basics, who wish to know more about how things work. I believe that there are simply too many students and patients whose questions cannot be answered because of the serious limitation of our knowledge. Tinkering is what most of us do as clinicians; therefore, we must try to improve the scientific base upon which we practice psychoanalysis or neurology. My greatest hope is that this book will improve the chances that individual researchers will seriously pursue interdisciplinary work in the area of mind and brain.

This book is the product of the input of many persons at different times during my life, although I alone take responsibility for the viewpoints presented. It is impossible to thank everyone, but I would like to mention Hoyt Alverson, Michael F. Basch, Mark Berger, David Dean Brockman, David Armstrong Brueckner,* Bertram Cohler, Timothy J. Crow, Takeo Doi, Dale Eickelman, Gordon and Paula Fuqua, John Gedo, Mark Gehrie, Arnold Goldberg, Martin Harrow, Tetsuya Iwasaki, Hiroyuki Kuramoto, Nils Lassen, Mark Levey, Samuel Lipton,* Nathaniel London,* Shin-Ichi Niwa, Itsuro Matsuo, Masahiro Nishijima, Nils Retterstohl, Leo Sadow, Ken Sakurai,* Nathan Schlessinger, Henry Seidenberg, Brenda Solomon, Harvey Strauss, Yasuhiko Taketomo, Marion and Paul Tolpin, Hiroshi Utena, D. Michael Vuckovich, Jerome Winer, Ernest Wolf, and Ryuji Yanai. Special thanks to Toby Troffkin, my copy editor, for her contribution to the logic and clarity of this book; to Eleanor Starke Kobrin, of The Analytic Press, for coordinating all phases of this project with grace; and to Paul E. Stepansky, Editor-in-Chief, for his wise input over the years.

*deceased

I submit that tension between science and faith should be resolved not in terms of elimination of reality, but in terms of synthesis.

Teilhard de Chardin

Science is not free of religion. . . . It is constantly involved in faith *that the ultimate truth will be uncovered—so magical ideas and expectations persist in scientific biases. Religious faith is invested in science and its results, and in its hope of improving the human situation. [But] attempts at complete objectivity are never successful.*

Roy R. Grinker, Sr.

Contents

Foreword—*John E. Gedo* xi

Preface xxi

1 Metaphor, Affect, and Arousal: How Interpretations Might Work 1

2 Psychoanalysis and the Two Cerebral Hemispheres—*with D. M. Vuckovich* 17

3 Brain Plasticity, Learning, and Psychoanalysis: Some Mechanisms of Integration and Coordination Within the Central Nervous System—*with D. M. Vuckovich* 43

4 The Prefrontal Cortex and Neural Control: The Brain's Systems for Judgment, Insight, and Selective Attention 83

5 The Hierarchical Developmental Model: Neural Control, Natural Language, and Recurrent Organization of the Brain 105

6 Integrating Sleep and Dream Research 121

7 Psychoanalysis, Nonverbal Communication, and the Integration of Touch (Contiguity), Vision (Similarity), and Hearing (Sonority) 145

8 The Transference Phenomenon: Possible Origin and Mechanisms 165

9 Management of the Transference: A Clinical Case Study 173

10 Psychological Development and the Changing Organization of the Brain 185

11 Some Notes on the Evolution of Language 201

12 Overview 219

References 225

Index 249

The Biology of Mind: A Foreword

John E. Gedo

Beyond therapeutic concerns, Sigmund Freud strove to develop psychoanalysis as the basic science of mental functions. Schooled as a physiological researcher by Brücke, Meynert, and Breuer, Freud was alert from the first to the necessity of finding a conceptual bridge between his observations of behavioral phenomena and the prevailing knowledge about the activities of the brain. As early as 1891, he postulated such a link in proposing the hypothesis of "psychic energy," a notion very much in the spirit of the scientific avant-garde of the day.

As a biologist, Freud was not merely a Darwinian, as Sulloway (1979) has cogently demonstrated; he was also a faithful adherent of the school of Helmholtz, one of the founders of which was his admired mentor, Brücke. The program of this scientific movement was the establishment of the life sciences on a solid basis of physics and chemistry—to put this in another way, the banishment of the last vestiges of vitalism from the realms of science. We may judge the seriousness of Freud's commitment to this enterprise from the pleas he was to make of C.G. Jung while the latter was in the process of defining his disagreements with psychoanalysis (see Gedo, 1983, chapter 13): Freud claimed that his theoretical proposals were an essential bulwark against the tendency of psychological systems to lapse into "occultism"—his pejorative designation for Platonic notions about the human soul.

We would therefore do well to look upon such Freudian concepts as psychic energy—the entire system of hypotheses he preferred to call "metapsychology"—as provisional proposals, made within a materialist framework. In this connection, it should be remembered that as late as 1895 Freud made a heroic, albeit unsuccessful, effort to ground psychology scientifically through a description in terms of putative processes in the nervous system. Freud had to abandon this "Project for a Scientific Psychology" because his understanding of neurophysiology, although completely up to date, was far from being equal to the task of undergirding psychology. The subsequent development of psychoanalytic theory on the basis of an entirely speculative metapsychology lacking in empirical referents was a desperate expedient—perhaps more indicative of Freud's need to anchor his thinking within the outward forms expected of a scientific enterprise than it was heuristically useful.

It might also be claimed that Freud's metapsychology maintained its plausibility for generations without obstructing progress in the clinical theory of psychoanalysis, although such progress requires major metapsychological revisions from time to time—theoretical revolutions such as Freud's own reformulation of his drive theories in 1920 and of his principal model of the mind in 1923, or, in the era following Freud's death, the evolution of "ego psychology" under the leadership of Heinz Hartmann (1964; see also Hartmann, Kris, and Loewenstein, 1964). As I have tried to demonstrate (Gedo, 1986), those analytic clinicians who made the boldest attempts to extend the applicability of psychoanalysis to populations beyond the boundaries of "neurosis" proper—for example, Ferenczi in the direction of syndromes of a more primitive variety, Melanie Klein in that of early childhood as well as of psychosis—were able to do so only by disregarding the requirement of theoretical coherence. In other words, over the years, more and more clinical observations were accumulated that would have been very difficult to reconcile with the prevalent theoretical schema had anyone made the effort to reestablish a unitary theory for psychoanalysis.

It was David Rapaport (1967) who first noted the resultant theoretical incoherence; he pointed out that the theories of object relations (which were proposed to systematize the clinical findings that did not seem to be explicable in terms of earlier schemata) could not be correlated with the drive theories that form the core of Freudian metapsychology.[1] Probably as a result of Rapaport's scientific rigor, it fell to his students to propose that the best way out of this quandary was to abandon the

[1]In later years, Loewald (1989) did make an effort to link these disparate theories by means of further speculative proposals; Kohut (1977) and Modell (1983), on the other hand, argued that psychoanalysis can afford to use several uncoordinated theoretical fragments. Although the point is defensible if we concern ourselves with clinical matters alone, such a policy would preclude the integration of psychoanalysis with psychology in general.

metapsychological paradigm (see Klein, 1976; Schafer, 1976; Gill and Holzman, 1976). Most of those radicals chose to abandon the biological pretensions of psychoanalysis altogether, confining their purview to the explication of those communications that are symbolically encoded—a psychoanalytic orientation often called "hermeneutic."

Not only have psychoanalytic hermeneuticists abandoned the Freudian ambition of establishing a science of mental functions, but also their aims fail to consider the influence of the preverbal stages of development on matters of clinical relevance. In this sense, they have disregarded Rapaport's (1967) insistence on the importance of the hierarchic organization of mental life, that is, that development must be understood as an epigenesis wherein the conditions of earlier phases are assimilated within those of later ones. Rubinstein (1976) was the first to point out that a psychology that accounts only for whatever is symbolically encoded cannot encompass the Freudian "unconscious." Although Rubinstein did not have access to enough information about brain functions to propose a neurophysiological alternative to the old metapsychology and to hermeneuticism, he devised a purely hypothetical "proto-neurophysiology" (Rubinstein, 1974) to demonstrate that when we know the functioning of the brain sufficiently well, we shall be able to understand the control of behavior on the basis of that knowledge without having to resort to hypothetical bridging concepts.[2]

More recently, an expanding cohort of scholars representing the next generation within psychoanalysis have mastered the flood of relevant information made available by the explosion of fruitful research in neurophysiology (see Hadley, 1985, 1989; Schwartz, 1987). Among them, Fred Levin has assumed the position of greatest scope, that of attempting to fulfill the program of Freud's 1895 "Project" and Rubenstein's (1974, 1977) work. Levin's effort is based on acceptance of the view that mental functions are hierarchically organized, that between the sensorimotor experience of the infant and the psychic organization of later life (characterized by the use of discursive symbols) there are intermediate modes of adaptation, based on communication by means of concrete signals or presentational symbols. In retrospect, it might be said that in the course of expectable human development language "enfolds" preverbal experience.

[2]In the meantime, most theoreticians have preserved a biological orientation. Those who have adopted a hierarchical view of mental life (see Gedo and Goldberg, 1973; Gedo, 1979, 1988) strive to conceptualize the automatic repetition of patterned biological experiences from the preverbal era in later phases of development; others continue to rely on the traditional vocabulary of metapsychology while admitting that they understand this as a series of metaphors. Regrettably, this position is lacking in scientific justification.

This process of learning naturally involves the momentous changes that constitute maturation of the nervous system.

Levin makes clear that the capacity of the brain to keep on changing, a capacity called "brain plasticity," underscores all subsequent learning, including the acquisition of those psychic functions that we usually understand by the term "analyzability." Another way to put this is that it has now been experimentally demonstrated that anything learned constitutes an anatomical change in the brain. It follows from this conclusion that the nervous system is best conceptualized as an organ of adaptation and that all psychological problems may fruitfully be viewed as learning disabilities. From this perspective, it is relevant to recall that the neural development of the prefrontal cortex is only "completed" around the age of 20, that is, when most people have learned to function as adults. If we accept the insight that adaptation is merely the dependent concomitant of neural control, it follows that it is mediated by the automatized decision-making processes of the brain—processes whereby information is used to produce goal directed behavior. The operational "grammar" that characterizes these events in the brain is not yet understood, but this is that automatic functioning of neuronal systems that psychoanalysis has called "the unconscious."

Like Freud before him, Levin postulates that psyche and soma are isomorphic, that the mind/body dualism proposed by Descartes (largely on religious grounds) is invalid, and that the vitalism that is hidden within contemporary mentalistic viewpoints merely denies our animal heritage. Reviewing man's evolutionary history, he stresses that all mammalians have brains that give priority to auditory and visual stimuli (rather than olfactory ones), a development paralleled by the adaptively advantageous attachment between the young and their mothers. Among extant animals, only the apes seem to have the capacity to use a sign language, a capacity inborn in the human neonate and utilized from the cradle to the grave. Vocalization apparently first developed in the early hominids and eventually evolved into the human capacity to articulate over 40 phonemes—a protolanguage originally lacking in symbolic connotations. Verbalization presumably began relatively recently; yet the development of human languages tends to mask the fact that our communications actually consist of an integrated assembly of verbal and nonverbal modes. It is the developmental line of these communicative channels that we are able to study simultaneously from the psychological and the somatic vantage points. In other words, information processing may be conceptualized from the viewpoint of either neuroscience or psychoanalysis.

Perhaps the most interesting of the findings about maturation of the brain highlighted by Levin is that connections between the two cerebral hemispheres by means of the corpus callosum are generally established around the age of three and a half, suggesting that the passage from the preoedipal to the oedipal period is dependent on the ability better to integrate the functions of the hemispheres—in the grossest of terms, those of primary and secondary processes. I am reminded of Stern's (1987) conjecture that entry into the psychic universe we call the Oedipus complex depends on the capacity to construct a narrative. At any rate, Levin points out that intrapsychic conflict becomes possible when different functional units of the brain (most likely the hemispheres) are in fact connected to each other—a conclusion congruent with my view, derived from clinical observations, that in regressions to archaic modes of functioning, mutually incompatible attitudes may coexist without conflict (see Gedo, 1988).

In agreement with Basch (1983), Levin views disavowal and repression as the means whereby thoughts are deprived of meaning by disconnecting affectivity from words and images. In this conception, repression is the process of blocking the input of the left cerebral hemisphere, whereas disavowal is the disconnection of right hemispheric input. Because of the advantages of communication by means of syntactically organized verbal codes, the left hemisphere becomes dominant (in right-handed persons) somewhere in the third year of life; this dominance facilitates gradual change toward the preferential use of repression as the typical defense, as the hierarchical model I proposed some years ago indicates (Gedo and Goldberg, 1973).

Although the bicameral mind implies that learning takes place in modules or subsystems that later undergo a process of integration (in 1973, Goldberg and I called these modules "nuclei of the self," and I later named the process of integrating them "self-organizational" [Gedo, 1979]), Levin makes clear that a core sense of self is formed well before the corpus callosum becomes fully functional. The weight of evidence suggests that the sense of self is originally a cerebellar function: the cerebellum forms maps of the body-in-space very early, so early, in fact, that it seems unlikely that the core sense of self could be experienced subjectively (i.e., consciously). Later these maps are duplicated in the central parietal cortex—I presume with accretions of remembered experience that transform these schemata into affectively charged motivational hierarchies (cf. Emde, 1983). Levin is explicit in postulating that we possess a sequence of models of "self-in-the-world"; he conceives of transference as a strategy of the brain to resort to one of the earlier of these models as an adaptive experiment. Consequently, he also believes that the hierarchical model of mind I have developed should be regarded as a reflection of the hierarchic organization of the brain.

The expanding repertory of communicative channels available to the infant (gestural, sonic, verbal, and syntactical) produces the progressive changes in brain organization we call psychological development. The acquisition of syntax reorganizes the brain most drastically; Levin reminds us that this achievement is not contingent on the capacity for speech: verbal communication is not necessarily superior to the sign language of the deaf.[3] In any case, it is the frontal cortex that is implicated in providing an overall organizer for behavior, and it is now known that this is the only area of the brain invariably implicated in the use of language. Levin reports some experimental evidence from Japan suggesting that the specific physical (prosodic) qualities and syntactical rules of the natural language one learns may influence the nature of the operational grammar used by the brain.

Instead of providing further details of Levin's review of the neuroscientific evidence he regards most relevant from a psychoanalytic vantage point, it may be more useful to consider whether his success in explicating all the major concepts of psychoanalysis in direct neurophysiological terms has implications beyond gaining impeccable scientific credentials for those concepts. My reading of this impressive effort to integrate two hitherto disparate realms of discourse is that it promises to alter profoundly both psychoanalytic theory *and* practice.

From the viewpoint of therapeutics, the cardinal implication of the new brain science is that treatment should be aimed at improving the information-processing skills available to the patient, a conclusion some authors have reached on clinical grounds (see Gardner, 1983; Gedo, 1988, Epilogue). In other words, the most important transaction in the process of analysis is the potential for the analysand to identify with the analyst's methods of data gathering and inference. Insofar as a therapeutic regression to conditions prevalent in childhood is a prerequisite for the emergence of the most relevant data, this requirement is probably promoted by the minimization of cerebellar input in the psychoanalytic situation. Analyzability may depend on the ability of higher centers to "manipulate" various cerebellar models of "self-in-the-world"; if these functions are not available in sufficient degree, one is able to respond to various contingencies only through enactments.

At the same time, the theoretical expectation behind the traditional technique of psychoanalysis, that of interpretation encoded in secondary-process terms, is revealed as ill grounded, for messages that rely entirely

---

[3]Fred Levin is one of the very few psychoanalysts capable of working in such a sign language; I believe that his experience in working with deaf patients was an important influence in alerting him to the importance of understanding thought processes in neurophysiological terms.

on discursive language are not likely to affect the more archaic layers of the hierarchy of schemata of the self. Levin advocates the analyst's use of metaphors to maximize communication with those levels of experience encoded in sensorimotor ways, in concrete signs, and in presentational symbols. His rationale is that metaphors are couched in linguistic symbols related to the various sensorimotor modalities. In cases where even such measures fail to establish adequate communication, it may be necessary to resort to methods that speak more directly to the right cerebral hemisphere—in right-handed persons, the seat of most of the dominant experience of the earliest years. Levin thus concurs with suggestions I have made in the past (Gedo, 1981, chapter 1; 1984, chapters 8 and 9) that communication by means of music and gesture may have to be employed—if you will, that the prosody of the analyst's speech may be as essential as its lexical content. He also points out that analysands may become blocked by neuronal gating within the brainstem; in these contingencies, internal processing (or the restoration of procedural memory) may be reestablished by means of "pump priming"—that is, the analyst's providing crucial associations (preferably encoded in nonverbal ways).[4]

It may be legitimate to summarize these therapeutic recommendations as an endorsement of the need for psychoanalysis to extend the theory of technique to cover measures "beyond interpretation" (see Gedo, 1979). Levin believes that it is the soothing effect of the analytic procedures—as Modell (1976) put it, or in Winnicott's terminology, analysis as a "holding environment", or its "empathic ambiance," as described by self psychologists—that reduces brainstem gating, thereby giving access to the earliest cerebellar schemata of the mind/body self. Another reason for noninterpretive interventions in analysis is the need to influence "habits," that is, behaviors that remain repetitive because they are kept largely detached from cortical control, within the corticostriatal system. *A propos,* we seem to be on the threshold of a neurophysiological explanation for the repetition compulsion. At any rate, it can no longer be maintained that psychopathology is primarily a matter of intrapsychic conflict. In the first place, much of it is embedded in character as a consequence of neurocognitive difficulties; second, whenever conflicts remain chronically unresolved, this condition must be understood as a failure of information processing—in other words, this state in itself should also be classified as a form of neurocognitive deficit. We have come full circle to a new appreciation—one that is better grounded in physiology—of Freud's concept of "actual neuroses." They are consequences of developmental lags, which we may now understand as failures of certain

---

[4]In these contingencies, it is access to cerebellar models of the archaic self that is most likely to be at issue.

crucial maturational processes in the brain, usually as a result of prior disorganizing experiences.

The psychoanalyst as Biologist of the Mind must oppose the recurrent temptation to believe that "in the beginning was the Word." The infant develops a sophisticated repertory of semiological functions before it learns the verbal symbols provided by the caretakers. In psychoanalysis, as in all of life, the verbal and nonverbal realms are closely linked. Nonetheless, Levin believes, as I do, that therapeutic success is more fundamentally dependent on the nonverbal components of the transaction than on its lexical content. These conclusions must not be taken to mean, however, that language competence is not essential. On the contrary, the acquisition of language provides the highest level of neural control in the prefrontal cortex; in this sense, cognitive functions are dependent on language. As Levin puts it, the operating system of the brain and the person's native language share certain rules: memories are coded nonsensorially in a "machine language" the brain develops parallel to the acquisition of linguistic competence.[5]

Nor would it be legitimate to jump to the conclusion that every instance of a syndrome that usually originates in the preoedipal (or even preverbal) era represents a direct homologue of archaic conditions. For example, Levin reports the finding that in sexual exhibitionists, the left cerebral hemisphere is unequipped—in the neurophysiological sense—to "police" the output of the right hemisphere. (Shades of the powerless rider on Freud's runaway horse!) Yet in the only two persons I have analyzed who suffered from this condition, the perversion supervened in adult life as a byproduct of states of exultation. I have no doubt that in such states of excessive arousal the output of the right hemisphere escaped the usual controls, but such controls were more than adequate in ordinary circumstances. In other words, the syndrome was, in these two cases, not the sequel of a developmental lag but an outcome of very unusual circumstances that few people would be prepared to cope with on the basis of early experience. I cite this cautionary tale in order to emphasize that a neurophysiological approach to behavior need not lead to abandonment of the vast array of valid clinical knowledge psychoanalysis has accumulated in the course of the past century; better understanding of the archaic, biological roots of our humanity should lead

[5]Levin makes note of the fact that the sign languages of the deaf seem to be closer to the basic linguistic code available to man than are languages using sonic symbols, as shown by the fact that it is easier to learn to use sign languages interchangeably. In other words, the crucial aspect of language acquisition is *not* verbalization.

instead to a more complex and nuanced view of the hierarchic adaptive possibilities of our behavioral repertory.

But emphasis should actually be placed on the obverse of this statement of reassurance to psychoanalysts long preoccupied with mental contents: the unfolding breakthrough toward a biology of mind promises soon to relegate hermeneutics to a secondary position in the analytic scheme of things and to focus primary attention on learning processes. The fruitful results of this coming revolution are incalculable.

REFERENCES

Basch, M. (1983), The perception of reality and the disavowal of meaning. *The Annual of Psychoanalysis,* 11:125–154. New York: International Universities Press.

Emde, R. (1983), The pre-representational self and its affective core. *The Psychoanalytic Study of the Child,* 38:165–192. New Haven, CT: Yale University Press.

Freud, S. (1891), Sketches for the "preliminary communication" of 1893. *Standard Edition,* 1:147. London: Hogarth Press, 1966.

_____ (1895), Project for a scientific psychology. *Standard Edition,* 1:283–391. London: Hogarth Press, 1966.

_____ (1920), Beyond the pleasure principle. *Standard Edition,* 18:3–66. London: Hogarth Press, 1955.

_____ (1923), The ego and the id. *Standard Edition,* 19:3–68. London: Hogarth Press, 1961.

Gardner, R. (1983), *Self-Inquiry.* Hillsdale, NJ: The Analytic Press, 1989.

Gedo, J. (1979), *Beyond Interpretation.* New York: International Universities Press.

_____ (1981), *Advances in Clinical Psychoanalysis.* New York: International Universities Press.

_____ (1983), *Portraits of the Artist.* Hillsdale, NJ: The Analytic Press, 1989.

_____ (1984), *Psychoanalysis and its Discontents.* New York: Guilford.

_____ (1986), *Conceptual Issues in Psychoanalysis.* Hillsdale, NJ: The Analytic Press.

_____ (1988), *The Mind in Disorder.* Hillsdale, NJ: The Analytic Press.

_____ Goldberg, A. (1973), *Models of the Mind.* Chicago: University of Chicago Press.

Gill, M. & Holzman, P., ed. (1976). *Psychol. Issues,* Monogr. 36. New York: International Universities Press.

Hadley, J. (1985), Attention, affect, and attachment. *Psychoanal. Contemp. Thought,* 8:529–50.

_____ (1989), The neurobiologyof motivational systems. In: *Psychoanalysis and Motivation,* J. Lichtenberg. Hillsdale, NJ: The Analytic Press, pp. 337–372.

Hartmann, H. (1964), *Essays in Ego Psychology.* New York: International Universities Press.

_____ Kris, E. & Loewenstein, R. (1964). Papers on psychoanalytic psychology. *Psychol. Issues,* Monogr. 14. New York: International Universities Press.

Klein, G. (1976), *Psychoanalytic Theory.* New York: International Universities Press.

Kohut, H. (1977), *The Restoration of the Self.* New York: International Universities Press.
Loewald, H. (1989), *Sublimation.* New Haven, CT: Yale University Press.
Modell, A. (1976), "The holding environment" and the therapeutic action of psychoanalysis. *J. Amer. Psychoanal. Assn.,* 24:285–308.
_____ (1983), The two contexts of the self. Presented at the 50th Anniversary Symposium, Boston Psychoanalytic Society and Institute (October 30).
Rapaport, D. (1967), *The Collected Papers of David Rapaport,* ed. M. Gill. New York: Basic Books.
Rubinstein, B. (1974), On the role of classificatory processes in mental functioning: Aspects of a psychoanalytic theoretical model. *Psychoanal. & Contemp. Science,* 3:101–185.
_____ (1976), On the possibility of a strictly clinical psychoanalytic theory: An essay in the philosophy of psychoanalysis. *Psychol. Issues,* Monogr. 36, pp. 229–264.
Schafer, R. (1976), *A New Language for Psychoanalysis.* New Haven, CT: Yale University Press.
Schwartz, A. (1987), Drives, affects, behavior and learning: Approaches to a psychobiology of emotion and to an integration of psychoanalytic and neurobiologic thought. *J. Amer. Psychoanal. Assn.,* 35:467–506.
Stern, D. (1987), The dialectic between the "interpersonal" and the "intrapsychic": With particular emphasis on the role of memory and representation. Presented at the 50th Anniversary Symposium of the Washington School of Psychiatry (April).
Sulloway, F. (1979), *Freud, Biologist of the Mind.* New York: Basic Books.

Preface

A number of books and articles have been written about the synthesis of psychoanalysis and neuroscience (see Levin, 1990). Where this book attempts to be distinctive, however, is in its effort to make *novel, specific,* and *detailed* correlations between psychological/psychoanalytic variables, on one hand, and neuroanatomical/neurophysiological considerations on the other. Most of the hypotheses stated in this text are testable and will be confirmed—or refuted—by the newer, noninvasive techniques for visualizing the brain in real time. The psychological and neurosciences have been converging as interest in the brain and in behavior has grown explosively during the last several decades. But the usual level of discourse describes this convergence in the most general of terms. *Mapping the Mind* examines specific psychoanalytic clinical observations in the light of some novel insights about brain mechanisms, and the reader is taken nearer the frontier of our knowledge of the brain.

The various chapters represent original contributions (some published earlier) to the scientific literature. Collecting this material in one place makes it possible to integrate insights across disciplines. To assist the reader, each chapter begins with a précis that places the chapter contents in overall context. In addition, chapter 12 further summarizes the subject matter and provides an overview.

Although there is no guarantee that what is serviceable today will be so for very long, for all theories are eventually replaced as deeper knowledge obtains, the contents of this volume represent my best understandings and insights. In addition, multiple perspectives and research domains have been consulted in the effort to find answers, since no individual or discipline could possibly master the complex questions being posed. If the result is at times complex, the reader must judge how much of this is my failure to elucidate clearly enough and how much is merely proof that humankind has yet to learn sufficiently about the workings of the brain.

Chapter 1 conceives of the regional cerebral blood flow studies of Lassen, Ingvar, and Skinhøj (1978) as relating to two contrasting mental states: in one, the subject's interest is mild, and his primary cortical association areas (for touch, vision, and hearing) activate sequentially; in the other, the subject's interest is peaked, and his primary cortical association areas for the three primary sensory modalities activate simultaneously. This observation, gleaned from Lassen et al.'s research report (but not at all their center of interest), suggests that coded into the analyst's transference interpretations are metaphors based on comparisons or equivalences between sensory experience in different modalities. This interdisciplinary perspective might be a key to understanding the power of "metaphorical" interpretations to stimulate sudden insight and simultaneous recollection through various bridging processes. The different intersensory integrating zones of the brain involved would seem to be areas of prime interest and importance for understanding such phenomena as insight, attention, remembering and forgetting, learning, and abstracting.

Chapter 2 pursues in detail one of the major insights of chapter 1, namely, the effect of interpretations on the bridging of cerebral hemispheres. A large body of research on brain asymmetry, including the insights of Galin (1974), Frick (1982), and others, is brought together with data from the field of clinical psychoanalysis to produce the hypothesis that "repression" and "disavowal" might be defined, respectively, as left-to-right and right-to-left interhemispheric communication blocks. This hypothesis represents a testable neuropsychiatric construct. While I was working on the ideas in chapter 2, Michael Franz Basch was studying similar issues involving the concise definition of the psychological defense of disavowal. Basch's and my concepts originally appeared in separate articles within the same volume of the *Annual* (see Basch, 1983), and I owe Basch much gratitude for his unselfish sharing of nascent ideas.

The work reported in the first two chapters on mechanisms of interpretation and psychological defense aroused my interest in understanding the basic mechanisms by which the brain captures experience (that is, learning what learning is, from an interdisciplinary perspective). However, to discuss learning fully, one needs to consider first the vast subject of brain "plasticity." This consideration is undertaken in chapter 3, which is organized around aspects of the clinical psychoanalytic situation that seem most explicable in terms of the vestibulocerebellar system. The work of Masao Itoh, formerly at the University of Tokyo and now at Riken Frontier Institute, shows how important for neural control (and hemispheric coordination) are the reflexive and voluntary feed-forward and feed-back control systems of the cerebellum. Itoh's research sheds light on how much learning depends on cerebellar activity. And a comprehensive review of learning research demonstrates that learning is the expression of plastic changes at various levels of organization within the brain. Psychoanalytic intervention unquestionably affects many of these levels, including this cortico-cerebellar system, and the cerebellar self-in-the-world model. It is the creation (as a developmental step) and internal manipulation of this mental model that allows thinking about objects (*gedanken* experiments) to occur, without our having to manipulate objects concretely in order to appreciate relationships.

Chapter 4 continues the investigation of neural control that began with the subject of the cortico-vestibulocerebellar system. However, now the field of purview extends to include four major interesting bodies of research: the cybernetics of the brain; the study of basic mechanisms of schizophrenia; the investigation of higher level brain activity using real-time scanning methodology; and the now familiar cerebellar mechanisms of neural control. It becomes clearer that adaptive decision making hinges on the coordination of at least three major subsystems of the brain (the prefrontal cortex, the cerebellum, and the basal ganglia). This chapter relates the psychological construct of adaptation, most closely associated with Hartmann and of great interest to psychoanalysts, to the subject of adaptive decision making, a neuroscientific, information-processing conception. Use is also made of some novel research reported in Japanese and translated by me. The reader will appreciate how "judgment," "selective attention," and "insight" can be defined in terms of the functioning of a core neural control system under the regulation of the prefrontal cortex. As indicated by the research of Ingvar, Kent, Niwa, Itoh, and Tsunoda, the prefrontal cortex not only plays a major role in neural control, but also makes a decisive contribution to human language capacity. Specifically, the work of Niwa and of Tsunoda lays the groundwork for a fuller appreciation of the complex relationship between the brain's operating system, native language, and psychological development.

Chapter 5, building on the insights of the previous chapters, focuses the earlier interest in brain organization on the developmental, hierarchical model of Gedo and Goldberg (as most recently modified by Gedo). This model is of considerable practical utility to psychoanalysis and has the added benefit of being supported by the evidence from infant observation studies, clinical psychoanalysis, and neuroscience. Questions are posed about the role of the timely exposure to one's language during infancy and childhood and the resultant tapping of one's fullest level of psychological development. Language, once assimilated, permanently and decisively alters brain organization. There is also reason to believe that our native language may contain recurrent, hierarchical elements that can be decoded as instructions to the brain's operating system; the function of the elements would be to rearrange the brain's functional organization and thus facilitate our solution of particular environmental problems. This would seem to be an important part of what we call "adaptation."

Chapter 6, originally written as an introduction to Max Stern's book *Repetition and Trauma: Toward a Teleonomic Theory of Psychoanalysis,* reviews research bridging sleep and dream studies, Stern suggested novel approaches to the problem of bridging neuroscientific and psychoanalytic perspectives, especially regarding the basis of psychological trauma. In the context of Stern's insights about the arrests in early psychological development that result in pavor nocturnus, I explore some further connections between nightmares, REM and non-REM sleep cycles, and the need for alternately "downloading" and "consolidating," or integrating, world views between the three or more major information-processing systems of the brain: the corticolimbic system, the cortico-vestibulocerebellar system, and the cortico-striatal system. In this manner, sleep phenomena and dream phenomena are related to each other by means of their possible relationship to the major learning subsystems of the brain.

Chapter 7 considers the phenomenon of nonverbal communication and suggests a scheme for classifying multiple communication, or langauge modes, each with its own brain-lateralization profile. This perspective is an expansion of the usual dichotomy between primary and secondary process. Gestural, prosodic, semantic, and syntactical modes appear to be the bare minimum necessary to cover the variously lateralized brain systems for communication or knowledge. After a brief series of clinical vignettes illustrative of the various communicative modes, the discussion proceeds to a major synthesis: The insights of Freud about "condensation" and "displacement" as fundamental mechanisms of "dreamwork," the conception of J.G. Fraser about two principles that explain all of myth and ritual ("contact"/"contagion" versus "similarity"), and Saussure and Jakobson's ideas that all of language is explained by two principles ("contiguity" and "similarity"), all seem to be the same two principles! Moreover, it seems possible that this set of rules itself reflects a more basic pattern of brain organization, namely, that the various sensory modalities of experience are integrated in a manner that establishes the function of "abstracting" as an aspect, or consequence, of the intersensory integration process. If the work of Fónagy and Hermann in linguistics (which deals with the significance of intonation or sonority as an important modality for affective information) were included, then there would be a fundamental brain integration along three primary axes: hearing (sonority), vision (similarity), and touch (contiguity). This chapter is rounded out by a consideration of the overall linguistic contribution to psychoanalysis.

Chapter 8 considers the mechanisms and possible adaptive significance of the transference phenomenon itself. Theory developed in chapter 6, explaining a possible information-processing basis for REM/non-REM cycles, is exploited and extended to what is conceptualized as a larger pattern that includes REM/non-REM cycles as well as transference/nontransference cycles. This overarching pattern reflects a fundamental pattern or strategy of the brain that provides added depth or dimensionality to self-reflection. The evolutionary benefit to the species of this improvement in the capacity for self-reflection is felt to be increased chances of survival.

Chapter 9 extends the previous considerations in chapter 7 (on nonverbal communication) and chapter 8 (on the transference phenomenon) by discussing a clinical case at length. The reader is provided with more details by which to judge my claim that the management of the transference is significantly facilitated by the recognition and sensitivity to nonverbal as well as verbal communications.

Chapters 9, 10, and 11 round out the book. Chapter 10 presents ideas on the relationship between psychological development and the changing organization of the brain. As noted earlier, the brain is devoted to decoding, processing, and encoding "communication." The brain responds to this experience by changing its organization in a process called learning, which involves a brain property called plasticity. And the permanent learning and plastic changes that occur over time are considered psychological development. Clearly, learning, plasticity, development, and communication all represent abstractions about related psychological processes. Chapter 10 is my attempt to consider the developmental sequence, more or less systematically, and generate a sense both of the order of change and of what these correlated changes might consist of in basic scientific terms. This effort means constructing two lines of developmental steps, one psychological and the other biological, that more or less connect with each other. The exercise is difficult, considering the limitations of our knowledge, and the large number of possible observations to match. But the effort, however tentative, seems important for future work in this area, which can only benefit from the errors that I make in creating such a developmental scheme.

Chapter 11 departs somewhat from the earlier considerations of language and communication and makes a brief statement on the evolution of language. The approach (in contrast to that of chapter 7) is phylogenetic rather than ontogenetic, and thus the perspective of language evolution is added to the other perspectives bearing on our analysis of the complexities of communicative modes in individuals. The hope is that the reader will better appreciate the difference between language and communication, as well as the concept that language and verbalization are not inextricably tied together.

Finally, chapter 12 is an overview of the journey taken in constructing the other chapters. It summarizes several core principles of brain organization observed over the previous chapters and makes some suggestions about needs for future research. It also invites others to follow in the path of finding better explanations for the phenomena observed. For psychoanalysis and neuroscience to remain teachable scientific entities requires more detailed knowledge of the underlying mechanisms of phenomena in each field. This means greater collaboration with colleagues in cognate disciplines. Interdisciplinary research offers us the best chance of finding those intersections that are the keys to genuine insight.

1

Metaphor, Affect, and Arousal

How Interpretations Might Work

PRÉCIS

Chapter 1 explores the power of words, particularly those words psychoanalysts call "interpretations." Growing up with impressions of the oratorical skill of people like Adlai Stevenson and Abraham Heschel, I became curious about what it is about certain forms of verbal expression that can excite the imagination. Man's emotional aliveness (his "affectivity"), expressed at times through words, appears to work through the redirection of attention and the alteration of critical brain thresholds for memory, insight, and learning.

The clinical material presented in this chapter is largely self-explanatory. The observations and speculations regarding neurophysiology are, however, more complex. Perhaps it will help some readers, therefore, to focus on the central concept of "bridging," which is approached from multiple perspectives. Bridging is important because it involves making connections, for example, between past and present experience, between personal experience (affect) and logical categories (cognition), between observations (sensation) and reminiscence (memory), or between conflicting tendencies or drives. "Such is the stuff as dreams [and therapy] are made on," and this chapter approaches such phenomena as bridging with the question: How does it work? The end result is a synthesis of the insights of Piaget, Freud, and contemporary neurophysiology (represented by Lassen, Ingvar, and Skinhoj), in which are described the decisive importance to the brain of the integration of information within the various primary sensory modalities: touch, vision, and hearing. One last point: this chapter contains a novel definition of metaphor, yet one that has never been formally challenged. I am defining metaphor as a verbal description that compares experience cross-modally, for example, that hearing (X) is like seeing (X) in a particular way.

Mr. D., a middle-aged, narcissistically disturbed social scientist, had a dream of a composer and a lyricist. The metaphorical transference interpretation was that the patient appreciated the extent to which his analyst had been able to acknowledge his feelings—as a lyricist puts words to the music of a composer. Especially useful had been his analyst's ability in the previous session to identify the patient's intense loneliness. Naming this feeling, which even the patient had failed to recognize, gave him a powerful sense that he was capable of understanding himself and of being understood by someone else, and more specifically by his analyst. He was *not* "psychotic," as he had accused his analyst of thinking of him and as he had secretly felt about himself.

It is my purpose here to explore one of the key issues of technique and theory, namely, the mechanisms of psychoanalytic interpretation, in which, I believe, metaphorical language plays a crucial role. If we can be explicit about our interpretive actions and words, we will learn more about what makes interpretations work. This might suggest something to us about our models of the mind, at least as systems for encoding (remembering) and (affectively) organizing experience.

To Freud (1914a), "the theory of repression is the cornerstone on which the whole of psycho-analysis rests" (p. 16). The various models of classical psychoanalytic theory can be viewed as systems for the organization of memories, which function as enduring psychic structure. Psychoanalysis is the process that undoes repression—often expressed by the patient as a "failure of memory"—by dealing with transference and resistance, "which emerge whenever an attempt is made to trace symptoms of a neurotic back to their sources in his past life . . . " (p. 16). But the questions remain: How precisely does one trace symptoms back to their source? And before, during, or after interpretations are given the patient, what actually happens that results in the opening up of the gates of memory and affect?

Strachey (1934) has suggested a general answer to the question of what makes an effective ("mutative") interpretation: (1) There is an effective *preparation* of the patient for the interpretation. In this "a portion of the patient's id-relationship to the analyst is made conscious by virtue of the latter's position as auxiliary superego . . . " (p. 283) This is a point I shall return to later. (2) The interpretation is aimed with *specificity,* "that is to say [it is] detailed and concrete" (p. 287). (3) There is the effect of the particular *language* of the interpretation. Here Strachey writes, for example, of the "blunting effect" of tacking onto interpretations an "ethnological parallel" or a "theoretical explanation." In essence, this chapter is an attempt to be still more specific about these insights of Strachey's.

The vastness of the general literature on interpretation precludes a comprehensive review here. To narrow the scope of this paper to manageable proportions, I would like to concentrate on that stage of analysis when transference is to the fore. I would also like to avoid, so far as possible, issues that relate to analytical style. The questions at hand are, rather, What is it in what each of us does, independent of our unique personality, that affects the analytic process in an effective way? What can be said about the state of the patient at the time an interpretation is made—that is, about the patient's readiness for experience, affect, insight, and so forth—as suggested by Strachey (1934)? How does the analyst recognize and/or create the patient's "receptivity" for interpretations? If some interpretations are more effective than others, there must be specific reasons for this. This chapter, then, represents an attempt to explore some aspects of the preparation, "specificity," and language of interpretations.

METAPHOR IN INTERPRETATION

Many psychoanalysts and psychoanalytically informed scholars have become interested in metaphor (Sharpe, 1940, 1950; Arlow, 1969; Lewin, 1969, 1970; Shapiro, 1971; Fine, Pollio, and Simpkinson, 1972; Reider, 1972; Rubinstein, 1972; Forest, 1973; Leavy, 1973; Rogers, 1973; Arieti, 1974; Litowitz, 1975; Rosen, 1977). However, these perspectives differ in emphasis significantly from my own. Most prominently, each deals with the role of metaphor in the patient's speech rather than in the analyst's. They also tend to treat metaphor as strictly indicative of relations within a semantic field (Rubinstein, 1972). They do, however, indicate how metaphorical language involves "switching" functions, ambiguity, multiplicity of meanings, symbolism, ease of comprehension, and thinking by similarities: thus, these perspectives recognize that metaphorical language is a complex behavior, relatively poorly understood but richly deserving of study.

Poets and literary artists speak to us in metaphor. They touch us emotionally. Perhaps one of the reasons for this effectiveness in the use of language is that by employing metaphors the artist is unwittingly tapping the richness of meanings that is a fundamental property of experience. All of man's mental activity is based on the ordering process of the central nervous system. But analysts and artists operate in fundamentally different ways, although there may be some similarities. No matter how artistic, the analyst's creative product (his analytic activity) is meant to be appreciated only for its value in facilitating an analytic process in the patient.

A reading of Freud's prose shows an active use of metaphorical language; there is, in fact, a separate index (of "analogies") in the *Standard Edition*. Metaphors constantly enriched Freud's writings and possibly his interpretations as well. Freud never commented on this in discussions of technique perhaps because its role was not apparent and because it was such an intrinsic part of his method that it seemed more stylistic than methodological.

In *Webster's New International Dictionary*, the Merriam Edition (1958), the word *metaphor* is defined as a derivative of the Greek *metaphor:* meta (beyond, over) and *pherein* (to bring, bear). It represents "the use of a word or phrase literally denoting one kind of object or idea in place of another by way of suggesting a likeness or analogy between them" (p. 1546). At a more fundamental level, at least part of what is carried or brought over from one sensory modality to another is the memory of the experience; that is, metaphors create bridges between sensory modalities. I shall return to this point in more detail toward the end of this chapter.

When a transference is interpreted in the language of an apt metaphor (apt for the patient), the evocative power seems enhanced and the analytic process is more likely to be accelerated. This is in part the result of the metaphor's having woven together for the patient a unique here-and-now experience, connecting past and present, but in a particular way. It is the particularity that is the subject of this chapter on technique. Additional strands of detailed recollection are thus suggested to the patient in an open-ended manner. (These strands themselves become bridges for, and catalysts of, further synthetic activity, a point that I will clarify later when I discuss the special and the general effects of the analyst's affectivity.) The memories of past and present can then be connected with a sense of vividness that convinces and allows for additional remembering and working through.

At this point it will be useful to consider some clinical examples so as to better illustrate the meaning of a metaphorical (transference) interpretation.

CLINICAL VIGNETTES

Mr. A., approaching 40, has an advanced degree in engineering. He began his treatment with a phallic-narcissistic character defense against his unresolved dependent longings and with a history of relationships with women in whom he would readily lose interest. We are now entering the fourth year of our collaboration. As his analysis has unfolded, we have understood the effects of a devastating series of early losses that involved an uncle (at age 5), father (at age 8), and a grandfather (at age 18) who had become a father-substitute. Against the backdrop of the years of our work the interpretations with metaphorical implications seem to have been the most effective.

Our most recent work involved cognition of the continuation of his mourning. Having had some of his clothes stolen from a laundry, Mr. A. spoke one day of his outraged embarrassment and his impulse to replace immediately the lost articles. He mused about having someone assist him in relation to this loss. I suggested that he needed a tailor and asked him if he knew of any way to mend the situation. This ambiguous metaphor was a reference to his major loss in childhood (his father was a tailor); to his recent loss (of clothes); and to myself in the transference as one who mends or helps him mend himself. He recalled with vivid details for the first time a particular garment his father had made for him just before his terminal illness and remembered his father giving it to him. With affect he continued with new details of the situation involving the later loss of the grandfather who had served *in loco patris* after his father's death. Shortly after this remembering and working through he was able to mobilize himself to obtain a job, after months of having been paralyzed in this regard.

Miss B., a 25-year-old prelingually deafened woman with an advanced degree, entered intensive psychotherapy approximately five years ago because of her feeling that in many ways her life had ground to a halt. I have described her elsewhere (Levin, 1977, 1980), especially with regard to the role of sign language in the defense transference. In a recent session she began by communicating an impulse to turn off the light near her "to save electricity." The previous session had dealt with her sexual feelings toward me and her boyfriend, difficult feelings for her to acknowledge. I reminded her of that session. She said, "You know what happens when the light is turned off" and moved on to more neutral subjects. After a while I brought her back to her remark and interpreted that she wished she could turn off her sexual feelings of excitement here just as easily as she could turn off my lamp. She agreed and went on to discuss in a sophisticated manner her situation with her lover and her plans for eventually resolving that situation favorably (in a way that sounded more realistic to me than some of her past plans). Toward the end of the session she said she had decided that the room light did not really need to be turned off. I commented that now she felt more in control of her excited, "electrical" (sexual) feelings and did not need to turn them off. She nodded and with a warm glow remembered for the first time how she had felt entering a "hearing" junior high school. She had previously been only in schools for the deaf. After a few weeks with hearing classmates she felt years behind in her work. For a while her father had helped her regain her equilibrium (as I just had), and before long she felt relatively caught up, although never fully comfortable.

THE PATIENT'S "LANGUAGE"

Over time analysts learn how a patient uses words, and over years we develop a vocabulary of his fantasies, feelings, impulses, inhibitions, dreams, and neuroses to which we call the patient's attention when we need to illustrate or organize a pattern near, but just out of, the patient's awareness. The reliable and apt recall of details from the patient's previous discourse is part of what convinces him that we hear, think, and care about him; he can now begin to experience himself as worth paying attention to, understanding, and caring about.

A patient's language probably also involves his particular blend of sensorimotor experience. Presumably, each of us has preferred combinations of sensations that excite, interest, or bore us. The verbal imagery that reaches us emotionally probably bears a relation to this mix. I shall return to this important point later.

Shannon and Wearer's (1949) communication theory suggests another critical element in any transfer of information—namely, the predictability of what is transmitted next. To the extent that the receiver of a message can predict what is coming, the value of the information derived approaches zero. The mathematics involves Markoff chains of a special class called "ergodic" processes. Metaphors thus surprise the listener in part because of their novelty. The improbable and therefore unexpected combination of ideas, sensory modalities, meanings, and so forth arouses the patient's interest; without it I do not believe synthetic activity can occur. With the use of metaphor the informational value of the message rises to a maximum (see Shannon's communication theory). If the reader has any doubts about this property of metaphors to surprise us, he should ask himself how often, when he is trying to communicate some difficult or important idea, he resorts quite spontaneously to a novel metaphor.

TRANSFERENCE AND THE THERAPEUTIC PROCESS

Valenstein (1961) quotes Anna Freud as suggesting that the psychoanalytic process is like a dinner, "where in Smorgasbord fashion a number of foods are set on the table for the choice of the diner. He selects foods in proportion and sequence according to his needs and the progress of the meal" (p. 315). Valenstein quotes Bibring to the effect that five psychotherapeutic principles are at work: suggestion, abreaction, the intentional use of the transference for support, insight through clarification, and insight through interpretation (p. 319). Strachey (1934) writes of "mutative interpretations" (p. 283) in which the patient is prepared, details are specified, and wording can help or hinder the effectiveness of the interpretation. In this role as "auxiliary superego" the analyst interprets an id impulse the object of which is the analyst himself. Thus, with the analyst's permission, the patient becomes aware of his impulses and defenses (in optimal dosage) and simultaneously of the contrast between the character of his feelings about the analyst and the real nature of the analyst; that is, the patient becomes aware of his archaic fantasied objects in the transference situation. Strachey thus regards transference as the patient's superego projected onto the analyst. What gets mutated is the superego that is reintrojected, based on a less harsh view of the analyst-as-archaic-imago and a remodeled view of the parent-as-archaic-imago.

Zetzel (1956) points out another view, that of Sterba and Bibring, in which an essential feature of the transference is an identification with the analyst. The superego is seen to have precursors, the development of which becomes relevant to the analystic work. Theoretically, there are a variety of ways to handle this situation of preoedipal problems. To Zetzel the rubric of the "therapeutic alliance" covers what others (Kohut, Gedo) represent as manifestations of archaic transference state. With this in mind, however, we can understand her feeling that the therapeutic process depends almost exclusively on transference interpretation; that is, analysis involves a regression in which there is a revival (reexperience) of primitive stages of development that can be interpreted as "here-and-now" experience.

Kohut (1966, 1971,1977) has painstakingly and explicitly expanded the definition of "transference" to include the continuation of an "unstructured" situation. In one of his earlier papers, Kohut (1959) indicates that "introspection in the case of narcissistic and borderline disturbances shows the analyst is not [merely] the screen for the projection of internal structures (transference), but expresses with the patient the direct continuation of an early reality that was too distant, too rejecting, or too unreliable to be transformed into solid psychological structures . . . " (p. 471).[1] In the more recent terminology of the self, the descriptive term *selfobject transference* has been used.

There is thus general agreement that the analyst's attention to transference through observation, introspection, and vicarious identifica-

---

[1]It seems that, theoretically at least, such arrests might also occur in every neurosis to some extent; why would one have trouble traversing the oedipal conflictual experience unless one already had had some difficulty, however minor, in a preoedipal stage of development?

tion (empathic means) and the interpretation of the transference are the *sine qua non* of an analysis. The question now is, Is there any contradiction between my preceding remarks on the details of interpretation—particularly on the use or role of metaphor—and the value of keeping one's eye principally on transference and its vicissitudes? I think not. The preceding remarks have been aimed at the detailed wording or structure of the interpretations, but I think it should be clear that these will primarily be transference interpretations. When the analyst, using a metaphorical style, tells the patient that he feels more whole because the analyst has been able to put words to his feelings in the same way a lyricist puts words to the music of a composer, this is a transference interpretation. The same is true when the analyst tells a patient, as in the foregoing example, that he is thinking of (re)finding a tailor (father) to mend himself. The discussion regarding metaphor must not be seen in any way as an attempt to propound a new theory in opposition to the one that has served us satisfactorily and that is confirmed daily in our work; that is, regressions set in motion by the analytic arrangement result in transference, based on fixations or arrests, that constitute current affective/cognitive experience and that we then interpret in a timely manner.

It seems to me that when metaphor can be found naturally and without artifice, it has many benefits: it treats with respect the patient's intelligence; it arouses the patient's interest, which is understood a priori to facilitate probable synthetic activity; it generally makes the transference interpretation easier to understand; and it has an ambiguity that allows for simultaneous relevance at multiple levels of experience and meaning. I would like to elaborate on this latter point in the following sections.

THE SPECIFIC EFFECT OF THE ANALYST'S AFFECTIVITY

The Freudian models of the mind, along with some neurological models that preceded them (e.g., those of John Hughlings Jackson in the previous century), implicitly contain the idea of memory as functional systems of the brain with hierarchical ordering. Whatever the current disagreements as to whether self is superordinate or subordinate mental structure, there seems to be general agreement that the brain is characterized by simultaneous multiple levels of functioning, with potential for the emergence of ontogenetically earlier modes of adaptive experiencing when later (higher) levels of the CNS become disabled for whatever reason.

Such a view is supported implicitly by clinicians who talk of addressing the patient's "observing ego" as opposed to his "experiencing ego"; it is also the explicit perspective of mathematicians interested in computers and the brain (for example, Von Neumann (1967) points out that unlike the brain, the digital computer operates linearly and at only one level).

The question remains how to relate the earlier perspectives with regard to the role of metaphor in transference interpretation to the prevailing view of the brain as such a hierarchical system. I believe that this can be accomplished relatively easily if we keep in mind that the diagrams that follow, as well as the descriptions associated with them, are meant only as maps to facilitate understanding—not as causal explanation. The fundamental problem is not in our models (which are understood to be temporary and tentative approximations, by definition) but in our tendencies to reify them (e.g., see Feldman and Toulmin, 1974–1975).

Interpretations that are effective provide bridges of various kinds. I have attempted to illustrate some of this bridging in Figures 1 and 2. In Figure 1, superimposed on a view of the left hemisphere of the brain is a triangle representing a linking up of three major sensory modalities: touch, hearing, and sight. Next to each letter are numbers in an inverted series to indicate the layering of functions according to a hierarchy within the framework of Piaget but entirely compatible with psychoanalysis (Basch, 1976a,). The details of this "horizontal" bridging appear in Figure 2 and are described in what follows.

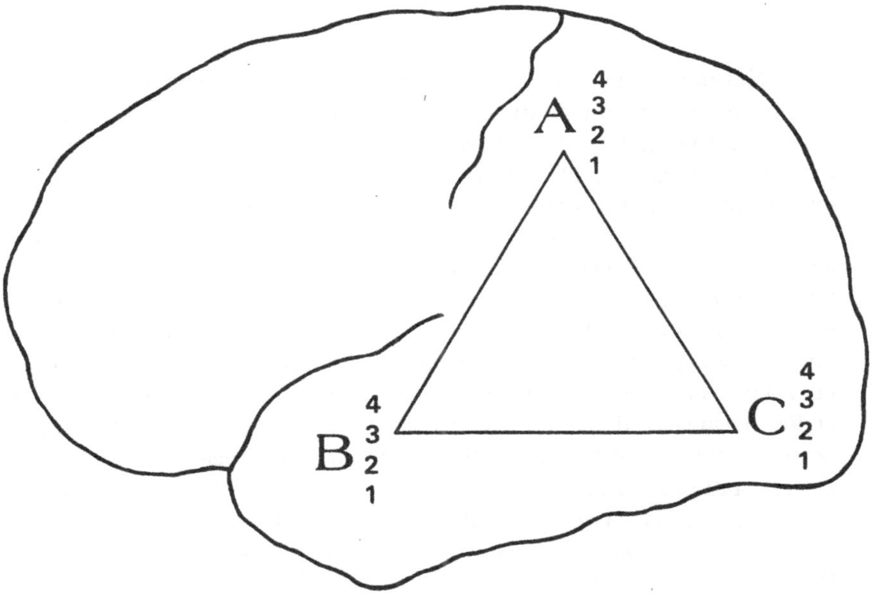

Figure 1

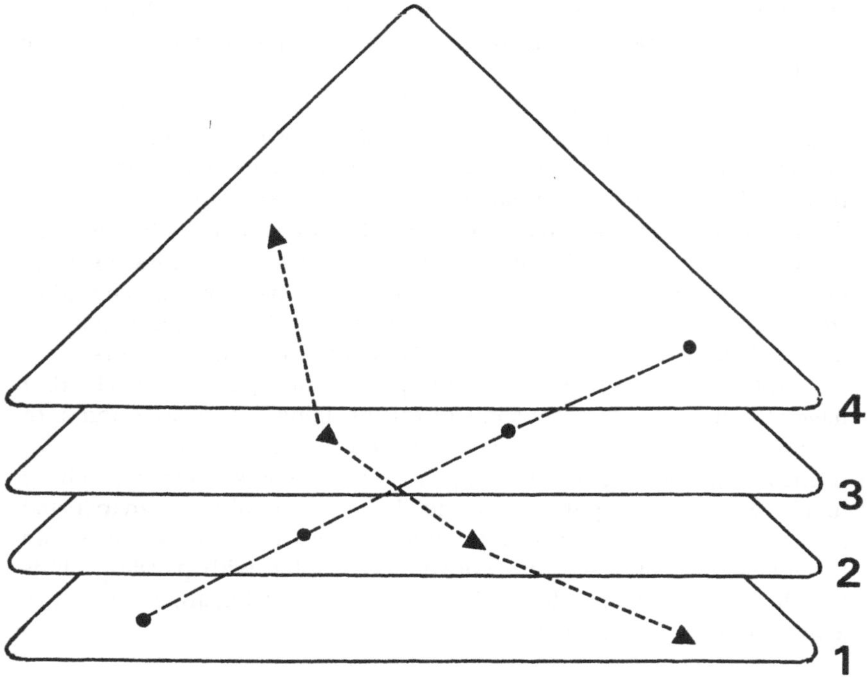

Figure 2

The dotted lines in Figure 2 portray "vertical" bridging within a particular sensory modality pattern complex. By this is meant some unique tying together of experience so as to involve in some way each of the four (arbitrary) levels indicated. At the lowest level (1), our metaphorical transference interpretation has touched on some important sensorimotor schema, which represents the encoding of autonomic affective experience organized into idiosyncratic rhythm patterns. At the next level (2), our interpretation is effective in the sense that its words serve to elicit evocative recall of still more complex and variegated memories of experience in which objects and subjects are named but in which essential characteristics of the classes named are superseded by accidental properties. Because of this latter fact, the recall is bound to be highly idiosyncratic or even disorganized to the extent that any recall evoked by our interpretation is influenced by this level of organization of experience. Finally, at the last two levels (3 and 4), presentational symbolism and discursive symbolism, respectively, are capable of being evoked. Although I am using Langer's (1967) terminology here, these levels coincide with Piaget's two highest levels. Time qualities (past/present) and essential rather than accidental properties of things are characteristic of the third level; objects can now be considered in terms of their properties. In the fourth level propositions of if/then, either/or, and hypothetical and deductive logic are now possible and will be manifested especially in secondary-process kinds of recall (discursive symbolism). In contrast, activation of the primary-process characteristic of the presentational symbolism of level (3) will be substantially harder to verbalize, to the extent that it contributes to the patient's response to our interpretation. However, it might be easier for us to decode, inasmuch as it may present itself as a purely visual memory (much like a dream) to which the patient and we can begin to associate.

The foregoing description is of course highly schematic, but it should communicate to some extent the ability of interpretations (when effective) to make us feel as though complex internal chain reactions of falling dominoes have started that run in several directions simultaneously and involve some reactivated memory processes that can possibly reorder our thinking and feeling. The nature of metaphor allows it some special effectivity in the direction of creating horizontal and vertical bridges of the type described above.[2] A good metaphor, when tied to a transference interpretation, is like a four-pronged plug that makes contact with each level of the patient's experience simultaneously. At the highest level (4) of conceptualization, verbal propositions are independent from sensory-modality quality (Basch, 1976a, b). At the lowest level (1) of sensorimotor schema are imprints of the unique blend of sensory-modality experience. As Basch (1976b) says of this period (the first 18 months of life):

> The parent's attitudes towards him are conveyed to the infant through tone of voice, rhythm of action, sureness of touch and other [sensory] signals which are to a greater or lesser extent not in the adult's awareness. The infant [in this stage] in responding to the kind and quality of the messages sent to him lays down the aforementioned [sensorimotor] action patterns that form the basis of his personality and are a response, so to speak, to what his parents are 'telling' him about himself and the world he has entered. This form of communication remains basic throughout life, though, for the most part, people continue to remain unaware of it [p. 9].

I interpret that by "this form of communication" Basch is referring to what we encode in our transactions with patients (without being aware

---

[2]Past and present are also connected, but general issues of time and timing have been deferred to a future essay. Time bridging could be referred to as "diagonal" bridging.

of doing so) that addresses itself to the sensorimotor experience of the infant. I agree with Basch that this in large part accounts for Freud's observation that the unconscious of one person can communicate with that of another. And my suggestion is that in part this is accomplished by reaching for metaphors with our patients—metaphors that will tap multiple levels of experience in ourselves and in them.

Metaphors cross modalities; they relate one sensation to another and the various hierarchical levels of experience to each other. In this manner metaphorical language contributes to the specific effects of the analyst's affectivity, that is, it seems to create the general affective arousal level that is required for synthetic activity to occur. Our continual problem is that the patient's changing moods require us to be flexible and not rigid or repetitive in our interpretations.

THE GENERAL EFFECT OF THE ANALYST'S AFFECTIVITY

In the foregoing sections I have attempted to outline schematically what I believe to be the importance of metaphor in making transference interpretations effective. When used naturally and without artifice, when not an end in themselves, and when apt (to the patient), metaphors will resonate with the highest and deepest layers of the patient's functional hierarchy of experience. I should now like to present some additional data from a related field that tends to confirm the observation that metaphor can function as a bridge, and particularly as a bridge across the lines of various sensory modalities.

Lassen, Ingvar, and Skinhøj (1978) report work that may be of interest to analysts. Using radioactive scanning techniques to study regional blood flow in the human brain in awake subjects, Lassen and his associates drew conclusions about the ongoing metabolic activity of the cerebral cortex. Their apparatus allows the visualization of the brain's surface cortical activity as it occurs! One of their most central observations relates to the role of the general level of arousal (of nonspecific pathways) of the brain. *When the arousal level is below a certain threshold of excitement, the patient's cortical activity appears to be limited to only one cortical (sensory) association area at a time.* Thus, for example, a relatively unaroused subject instructed to imagine himself listening to a Beethoven sonata will activate only the auditory associative cortex in the temporal lobe. However, if a threshold of interest is exceeded, the brain becomes activated *as a whole,* and (important for this discussion) the various associative cortical (and presumably also the subcortical) parts of the brain come into communication with each other.

Lassen and his associates assume—I think correctly—that learning about the world (and, one would have to add, the self) can then occur; that is, synthetic activity of the brain would appear to be a function of this general level of arousal of the brain's nonspecific pathways. It seems to me we have here a general explanation as to how our interpretations might work (Lassen, personal communication). We could say, preliminarily, that the general effect of the analyst's affectivity[3] is to help bring about this state of arousal of the brain as a whole (or to recognize when it has occurred). As already stated, it seems likely that synthetic activity occurs when the patient's mind is aroused and in communication with itself. Moreover, as noted in the preceding section, this general state of arousal appears to be facilitated by specific interpretations and more especially by linguistically coded, metaphorical interpretations that tap multiple levels of meaning, including (affect-laden) sensorimotor schema.

THE ROLE OF SENSORY MODALITIES

An important aspect of synthetic activity as it is understood by Lassen and his associates is the functional isolation of the various sensory modalities according to the topography of the cerebral cortex. Memories for each sensory modality are in all likelihood not neatly localizable topographically in any simple manner but seem to have some functional dispersal.

The role of sensory modality in the organization of memory was implicit, and often explicit, in Freud's early theoretical thinking. (For example, the rungs in the reflex-arc model were meant to be the various sensory modalities). It even played a role in Freud's thinking about one of the basic subjects of this paper, namely, repression, which Freud (1897) defined as follows in a letter to Fliess: "To put it crudely, the current memory stinks just as an actual object stinks; and just as we turn away our sense organ (the head and nose) in disgust, so do our preconsciousness and our conscious sense turn away from the memory. This is *repression*" (p. 269). In *Studies on Hysteria* (Breuer and Freud, 1893–1895), Freud described the case of Lucy R.,who was tormented by subjective sensations of smelling "burnt pudding." Freud states that in order to begin to unravel its meanings, "I only needed to assume that a smell of burnt pudding had actually occurred in the experience which had operated as a trauma" (p. 107). References to the complex fabric of sensory modalities as mnemic organizers occur repeatedly throughout Freud's work and apparently played an important role in his own organization of psychic data.

---

[3]I refer here to affectivity rather than simply to interpretive technique, because I believe, along with Gedo (1978), that there is no communication without affect.

Without our realizing it, in our work with patients we take advantage of the principle that memory is organized along the lines of the different sensory modalities. Again, as follows from the work of Lassen and his associates, we must in some way, through our interpretations, be fostering a process in which a general state of arousal occurs. Since this would appear to involve bringing the various psychical agencies into communication with each other (bringing, for example, the memories that are organized according to various sensory modalities into functional connection with each other), our interpretive activity might be seen as inviting a transfer across sensory modalities.[4] How might this occur?

A tentative answer is that *one of the crucial elements in an effective interpretation is metaphor. Here metaphorical language may even serve as a functional bridge between various psychical agencies that might not be otherwise connected at the time, and in a manner that would allow transfer and creative synthesis of information.* Some of our best (most effective) interpretations might be those that have clear metaphorical aspects, at times implicitly but perhaps preferably explicitly; and our use of metaphor in these cases will not be accidental or incidental but will probably be a reflection of some decisive role of metaphorical language in the coding and in the transfer of information within the mind.

The descriptions of "functional bridging" given in previous paragraphs use some linguistically coded means (e.g., metaphor) that focus on the need to understand better the internal communicative and relational aspects of mental systems. Metaphor in this chapter is used both concretely and abstractly to denote the complex subject of the hierarchical organization of the mind and its regulation. Although it is not discussed in this paper, it will be obvious to some readers that functionally organized mental structures may work in concert as well as in conflict (see chapter 2).

---

[4]There is a fascinating reference to what may be a similar or related phenomenon-namely, *transfert* in the sense of hysterical transferring of sensibility, which appears in the preface to Freud's (1888) translation of Bernheim's book on hypnosis. The transfert is from one side of the body to another. Freud sees this as "proving the suggestive origin of hysterical symptoms" (p. 79). Later he states, "It is merely an exaggeration of a relation which is normally present between symmetrical parts of the body." This raises the question of the role of the different functions of the two hemispheres of the brain, a subject about which much has been written, and how this splitting of function between the hemispheres is related to the subject of transfer of sensory experience across modalities, transfer of learning, etc. I could not hope to discuss these subjects comprehensively in this chapter; they are richly deserving of a separate presentation.

SUMMARY

In this chapter I have attempted to begin to describe the complex subject of the role of metaphor in the analyst's interpretations; the original insights of Strachey about the specific mechanisms of effective interpretation are pursued. When apt, metaphors appear to play a crucial role for the patient, first of all in producing in him a state of general psychical arousal that allows for synthetic activity. (I have characterized this threshold shift in terms of the general effect of the analyst's affectivity.) This is part of the meaning of Strachey's point that interpretations require a preparation. Second, the metaphorical language of the analyst's transference interpretations has precise specificity in this arousal of the patient's interest by creating specific or idiosyncratic "vertical," "horizontal," and time-dimensional ("diagonal") bridges within his mind that tap his unique hierarchical ordering system of encoded experience. (This detailed bridging activity, in all its complexity, constitutes the specific effect of the analyst's affectivity, much as a particular key works in a specific lock.) Metaphors thus cross sensory modalities and address the patient in a manner that respects his intelligence and that is concrete and abstract, comprehensible as well as integrating. Their novelty evokes arousal; their familiarity evokes synthesis; and their relation to the transference makes the whole thing go.

2

Psychoanalysis and the Two Cerebral Hemispheres

with D. M. Vuckovich

PRÉCIS

The previous chapter considered various kinds of bridging within the brain but left for Chapter 2 a fuller examination of the subject of the bridging of the two great cerebral hemispheres. Much has been written about how these two brains differ but less on how they actually collaborate. This chapter concentrates on the need for hemispheric collaboration and the significance of blocks between the hemispheres. Years ago I saw a film showing a patient with a severed corpus callosum, the central band that connects the hemispheres. The striking aspect of this patient was that when he was asked to use his hands to put together a puzzle, his hands (each under the control of a different hemisphere, which was not in communication with the other side) behaved as though they were the hands of two people who could not collaborate with each other at all! At one point the patient's right hand suddenly pulled the puzzle away from the left hand and refused to turn it over.

The central idea of this chapter is that it is possible to take what psychoanalysts call defense and what neuroscientists call interhemispheric communication and relate these to each other in some interesting ways. My theory (testable with modern noninvasive techniques for brain visualization) is that what we call repression (forgetting, especially of highly personal experience) and disavowal (downplaying the emotional significance of experience) are left to right and right to left blocks, respectively, of the flow of information between the cerebral hemispheres. This theory occurred to me while I was listening to a lecture by Nathan Schlessinger on the subject of follow-up studies in psychoanalysis and was thinking of empirical evidence that an emotional conflict is something that is resolved in psychoanalysis through a process in which competing tendencies or impulses become more rapidly cycled and resolved (rather than demolished). I was also recalling an article I had read in a technical journal about meandering rivers, namely, that rivers in the northern hemisphere erode at the bottom especially on the right side (meaning to the right of the direction of flow) whereas the reverse is true for rivers in the southern hemisphere (a phenomenon that Einstein had explained to the Prussian Academy in 1926). From alternating between the two meanings of "hemisphere" (Northern vs. Southern, and right versus left hemisphere of the brain), I considered the idea that conflicts might be dealt with by controlling the cycling of information between the cerebral hemispheres. I then shared this information with my collaborator, D. M. Vuckovich, who immediately recognized the utility of such a theory for explaining something in the neurological literature that had never been satisfactorily explained—namely, certain cases that are exceptions to the rule of Pitres (as reported by Minkowski), which describes how multilingual patients with strokes recover language (they usually recover first the language they were using at the time of the stroke). Thus the conception of Chapter 2 began with some interesting applications of a core theory to two quite different bodies of data. This sort of experience—that is, when a theory seems correct from two or more different simultaneously applied perspectives—has always made me more confident of the results. Of course, the experienced reader will appreciate that this is exactly the advantage of interdisciplinary research.

This chapter attempts to review recent neurological knowledge of the two cerebral hemispheres and describe the general significance to psychoanalysis of the brain's bicameral structure. The adjective *bicameral* is used throughout this chapter to mean the functional specialization and any associated anatomical asymmetries of the system of two cerebral hemispheres. The reader will be able to follow from the context whether the referent is brain structure (anatomy), brain function (physiology or psychodynamics), or both. Detailed delineation of the nature of cortical brain asymmetries are presented under the heading of Hemispheric Stereotypes. In our opinion, the growing interest in and knowledge of the brain have not yet been adequately explored or significantly enough integrated into psychoanalysis (Lehtonen, 1980).

The history of the modern study of brain asymmetry starts with the unpublished findings of Marc Dax (Gibson, 1962; Springer and Deutsch, 1981), an obscure French general practitioner who reported to his local medical society in 1836 on more than 40 patients he saw for aphasia.[1] Dax noted evidence of damage to the left hemisphere in all cases, whereas in no case did he find evidence of damage to the right hemisphere alone. The speech function was thus connected with the left hemisphere for the first time. It was later, after he learned of Dax's research, that Broca understood this relation for the first time (Gibson, 1962).

Brown (1969, 1972, 1974) has reviewed the investigation of aphasia from Gall (Gall and Spurzheim, 1810–1819) to Wernicke (1874) and beyond, a history that blends into the modern pioneering work on brain asymmetry by Sperry (1968, 1970), Gazzaniga (1970; Gazzaniga and Hillyard, 1971), Geschwind (1964, 1968, 1972), and Brown (1969). These split-brain studies were presaged by the still earlier work of Freud (1891a), Lashley (1937, 1951), Penfield (Penfield and Roberts, 1959), and Eccles (1978), as well as that of many others, on the critical problem of localization of brain function (see especially Gibson, 1962). According to Locke (1977), belief has fluctuated from holding (overpessimistically) that "efforts to study the relation between mind and brain . . . [are] . . . unsatisfactory because of the difficulty of objectifying the phenomena of mind" (p. 75) to asserting (overly optimistically), as Lashley (1951) does, that "the phenomena of behavior and of mind are ultimately describable in the concepts of the mathematical and physical sciences" (p. 121).

Although some modern reviews of brain asymmetry studies do exist (Galin, 1974; Levy, 1974; Meyersburg and Post, 1979; Walker, 1980; Wexler, 1980; Springer and Deutsch, 1981), they do not uniformly take into account the special interests of psychoanalysis as opposed to general psychiatry. We shall make an attempt to improve this situation in this chapter, although the reader will appreciate that a comprehensive review of such a complex subject as bicamerality of the brain and its relation to psychoanalysis is probably better suited to a monograph. Nevertheless, we shall address what we feel to be some of the central areas within this larger domain. This chapter attempts to complement work in each discipline and to contribute to the identification of isomorphic principles[2] for

---

[1]It is probable that the majority of Dax's cases were male, since we now know that females are significantly less lateralized than males and more likely to recover language function after brain damage. The entire question of sexuality and brain asymmetry, however, is insufficiently clear to be included in this review (see McGlone, 1978).

[2]By "isomorphic principles" or "isomorphism" we are referring to the specific viewpoint of general systems theory (von Bertalanffy, 1968) in which laws identical in structure are discoverable in intrinsically different fields. Unlike scientifically worthless analogies (superficial similarities) or scientifically important

neurology and psychiatry.

The cerebral hemispheres (which with their asymmetrical properties are in essence two brains), are capable of sharing internally and externally generated experiences. The information at the linguistic or ideational level evokes the image of the "liaison brain" (Popper and Eccles, 1977). Recent studies have speculated (probably in an oversimplification) that the dominant hemisphere is the site of self-consciousness (Eccles, 1978).[3] No one knows exactly why our bicameral mind evolved. Demonstrated most clearly in the language potential and handedness of homo sapiens, the asymmetry of the brain allows an economical utilization of cerebral substrate without unnecessary duplication of function. With a steering mechanism packaged in this manner (in the form of a double brain), the organism would seem to have an evolutionary advantage. Later we shall speculate in more detail about the nature of the evolutionary advantages conferred by the development of brain asymmetry (language and handedness) that, according to the best archaeological evidence, occurred two million years ago in the ancestor of man known as Australopithecus.

Cortical morphological differences between the two hemispheres are present before birth and have also been observed in the temporal lobes of fetuses at six months of age (Wada and Davis, 1977). The early establishment of cerebral asymmetry is further attested to by the differential responses to auditory and visual evoked responses of what are to be the dominant and nondominant hemispheres respectively. According to Levy (1974), approximately 89 percent of the population is phenotypically right-handed, and 99.67 percent of this group is left-hemisphere language dominant. Of the 11 percent of phenotypic sinistrals, 56 percent have left language dominance and 44 percent right language dominance. This asymmetry in language functions has profound implications for the nature of human experience and for the special study of aphasia, which we shall present in more detail.

Optimal interhemispheric communication is established through

---

homologies (where, beyond mere similarity, there is enough identicalness to assist in the formulation of valid models), isomorphisms between fields entails an *explanation* of the phenomena involved in the sense of "the statement of specific conditions and laws that are valid for an individual object or for a class of objects" (p. 85).

[3]Of course psychoanalysis, following Freud's lead, has focused not on consciousness but on the functionally more important continuous fabric of the mind's operations. The study of unconscious mechanisms is the study of the clinical correlate of neuronal systems, most of which function automatically (i.e., out of our awareness).

the corpus callosum and through cerebellar pathways, as well as by the hippocampal, anterior, posterior, and other commissures; this linking up into a system with complex connections is of fundamental importance in terms of the brain's capabilities to adapt to environmental changes, to encode information that can subsequently be retrieved, and generally to profit from experience. Only in the past eyeblink of human evolution has the bicameral brain, through such sciences as neurology and psychoanalytic psychology, begun to study itself.

As stated earlier, cerebral lateralities are established in the prenatal period, and the asymmetrical characteristics and the development of inborn capabilities for sharing by the two hemispheres are an ongoing process. Commissural (interhemispheric) transmission begins in early life and accelerates, being clearly defined first at three and a half years and completed by adolescence (Wexler, 1980). Thus, the beginning of the oedipal phase, a psychological and neuroanatomical watershed in development, coincides with the onset of the ability (or inability) of the hemispheres to integrate their activities.[4]

Theoretically, it is possible for the two hemispheres to be either not connected at all or suboptimally connected; in practice, the former situation corresponds with the syndrome of congenital absence of a corpus callosum. When this occurs, the two hemispheres develop the ability to function independently, incorporating skills in each hemisphere that are not usually there. This somewhat puzzling development requires detailed explanation, unfortunately not yet available, of the relationship between the potential plasticity or equipotentiality of cortical cells in modular arrangement and their normal tendency to differentiate and specialize with resulting loss of plasticity.[5]

---

[4]There is disagreement in psychoanalysis over whether the oedipal phase is necessarily connected with turmoil (see Kohut, 1977). It is also interesting to speculate that there may be significant individual differences in the exact time of onset or pattern of myelinization of the interhemispheric connections, with resulting differences in the kind of "mind" that is applied to the problems of the oedipal age (bihemispheric or lateralized). Unfortunately, we cannot fully clarify, at present, how these two different situations might be discriminated from each other experientially and/or observationally except to note that lateralized activities will show the special stamp of the hemisphere that is especially involved (see the section Hemispheric Stereotypes).

[5]The subject of the brain's plasticity seems to be crucial to the practice of psychoanalysis because it plays a decisive role in analyzability. There is even reason to believe that anatomical changes in the brain may occur secondary to analysis. The interested reader is referred to an article describing experiments in which sexual hormones and steroids (variables that fluctuate along with mood during psychoanalysis) produced dendritic growth in the adult brain of a related

There are also problems in assuming that the two hemispheres really develop with complete autonomy, since there is no precise or complete knowledge regarding all potential interhemispheric connections, such as of the hipoocampal commissure or linkages within the vestibulocerebellar system. As another example of this complexity, it is known that in those human beings born with agenesis of the corpus callosum, the anterior commissure is correspondingly enlarged (Kolb and Whinshaw, 1980) and, by implication, compensatory.

METHODOLOGICAL PROBLEMS

A number of methodological problems arise from our attempts to relate neurological to psychoanalytic concepts. As reviewed by Hill (1981), the major one stems from the continuing philosophical debate between psychological dualism/vitalism on one hand, and modern materialism on the other. The former is represented by the line of thought of Plato, St. Augustine, Descartes, Leibniz, Spinoza, Eccles, and Popper; the latter, by the theories of Hippocrates, Galen, Aristotle, Democritus, Pelagius, Hobbes, Armstrong, and Freud. Although the central debate is not easy to summarize, it seems to hinge now on whether or not one believes that *all* aspects of mental life are capable of being understood in terms of brain mechanisms. Thus, for Popper and Eccles, although there is a liaison between the brain and the conscious self, the relationship is not a mechanical or a complete one; the computer/brain is understood to be helpless without a programmer, the self. Eccles even postulates the immortality of this immaterial entity (the conscious self).

In addition to the debate over what neuropsychiatric correlations might mean (e.g., regarding causality), there are technical/procedural differences as to how one should go about studying these correlations. In fact, the large number of approaches to interdisciplinary work staggers the imagination. Table 1 lists a very small sample of such studies. For a detailed general review of the issues involved in choosing among these approaches, the reader is referred to Kolb and Winshaw(1980).

Still another methodological problem is how one establishes the truth of suggestive correlations. In addition, there is the problem of defining the range of applicability of any new models. For example, later in this chapter we describe a new way of defining conflict as relating to weak connections between the two hemispheres. The question will arise whether or not the clinical correlations we make are indeed there to make and whether, if they are true (causal), it is possible to distinguish between this new kind of "conflict" and conflict as understood in terms of the

species (De Voogd and Nottebohm, 1981). A full discussion of this fascinating subject, however, would take us beyond the subject at hand.

TABLE 1
Interdiciplinary Approaches to Neurology and Psychiatry

Subjects Studies
The neuropsychiatric examination
Neuropsychological testing
Clinical-pathological studies of brain lesions: aphasia, infection, vascular disease, tumors, trauma, etc.
Comparative neuroanatomy and embryology of primates and humans
Left-right brain studies: handedness, dichotic listening, tachistoscopic studies, split-brain studies
Language studies in humans and subhuman primates
EEG research, evoked potentials, brainstem evoked response testing, magnetic resonance imaging, isotopic scanning, and positron-emission studies of the awake brain
Neurotransmitter and neuropsychopharmacology studies
Electrical stimulation of the brain (as in the work of Penfield or Delgado)
Studies of the patterns of recovery and sparing of function after brain damage
Dyslexia and stuttering research
Artificial intelligence

traditional psychoanalytic models (topographical and structural). As for the second question, we suggest a way to make this distinction, although it does not offer anything definitive. As for the first question, however, we must admit that we cannot prove the new definition of conflict sufficiently to satisfy ourselves and the critical reader that it is true or useful absolutely. And yet we believe that examples that we introduce toward the end of this chapter will demonstrate the relative utility of our viewpoint and make it probable that our correlation (between conflict, as represented in the phenomena of repression and disavowal, and hemisphere disconnectedness) is correct. To paraphrase Piaget (1971), one is no longer dealing with biology or psychology vis-à-vis epistemology but with the convergence between these ways of thinking (i.e., we are looking for isomorphisms).

Hartmann, Kris, and Loewenstein (1953) wrote of the distinction between Freud's pre-and post-1937 view of the ego, which involved a revision taking into account the ego's hereditary predispositions (which Freud had previously ascribed only to the id) and the physiological apparatus at the ego's disposal. According to Hartmann and his associates, this represented a clear broadening of psychoanalysis.

This aspect of Hartmann, Kris, and Loewenstein's views is also well articulated in another article (Frick, 1982) that, like our own, attempts to integrate some psychoanalytic and neurophysiological viewpoints.

Frick suggests a hierarchy of ego functions that may be seen in the influence of the cerebellum (and its vestibular pathways), a point of considerable importance. At the brainstem level there is a vestibulocerebellar system (VCS) influence on posture and integrated action. At the mesolimbic level there is the possibility of the VCS modulating drives and aiding the integration of the sense of self and consolidating object relations. And at the cortical level the VCS may organize secondary-process thinking and/or "contribute to coordinated activity between left and right hemispheres" (p. 117).

HEMISPHERIC STEREOTYPES AND THE CASE OF APHASIA

What, then, are some of the relative differences between the hemispheres? The right hemisphere is a leading part of the brain's systems concerned with visuomotor and spatial skills, facial perceptions, the prosody (metrical composition) of language, and time synthesis. It is also implicated in musical activities, particularly those involving tonal recognition and reproduction (i.e., without direct correlation with musical scores). It appears from studying the effects of right-sided brain lesions that the right hemisphere has a significant affective function, for example, in the prosody of speech (intonation), since its damage interferes with expressive speech as well as with the recognition of speech nuances (Tucker, Watson, and Heilman, 1977). The extent of right-hemisphere language is variable. In some cases it appears able to generate speech; in others there is primarily comprehension of words and little expressive ability (Sidtis et al., 1981). In spite of commonplace statements that language regulation is primarily a left-brain phenomenon, there is evidence that the right brain contributes significantly to certain aspects of word fluency, prosody, and verbal memory (Tucker, Watson, and Heilman, 1977). Actually, several speech centers may exist—usually in the same hemisphere, but occasionally in the opposite hemisphere—which explains why sometimes we see recovery of language in those whose anatomical lesions tend to militate against such a possibility. Thus, the actual situation of hemispheric differences is considerably more complex than one might conclude from reading most articles about simple dichotomies.[6] Finally, evidence exists that the right hemisphere (left side

[6]An example of this complexity is the particularly interesting question of whether or not so-called split consciousness is really possible (see LeDoux, Wilson, and Gazzaniga, 1977). Since this chapter concentrates primarily on unconscious or automatic neuropsychiatric mechanisms, about which a great deal more is known, we have elected not to discuss this question. We would, however, state

of the body) is superior in processing certain kinds of sensory input, such as tactile stimulation (Moreau and Milner, 1981).

Freud wrote extensively on aphasia, which he described as a functional cerebral-disconnection syndrome. To explicate the specific issue of localization of brain function (as well as the general problem of relating specific neurological concepts to those of psychoanalysis),we will try to describe in more detail some aspects of Freud's (1891b) book *On Aphasia*, which does not appear in its complete form in the *Standard Edition*.

Reviewing Freud's book, Brown (1975) describes the ideas of association disturbance as a "daring—indeed, revolutionary" (p. 246) refutation of the narrow anatomical model of localization (centers and pathways) of Wernicke and Lichtheim that was then prevalent. According to Freud (1891b):

> The central apparatus of speech [is] . . . a continuous cortical region occupying space between the terminations of the optic and acoustic nerves and the areas of the cranial and certain peripheral motor nerves in the left hemisphere . . . [However] we have refused to localize the psychic elements of the speech process in specified areas within this region [p. 67].

In this "speech zone" or "speech apparatus," according to Freud, *"all aphasias originate in interruption of associations"* (p. 67; italics added), that is, of conduction. To Freud, the underlying principle was "purely functional" (p. 104), that is, no longer strictly anatomical or topographic. He is worth quoting in more detail:

> Considering the tendency of earlier medical periods to localize whole mental faculties, such as are defined in psychological terminology, in certain areas of the brain, it was bound to appear as a great advance when Wernicke declared that only the simplest psychic elements, i.e., the various sensory perceptions, could be localized in the cortex, the areas concerned being those of the central terminations of the sensory nerves. But does one not in principle make the same mistake, irrespective of whether one tries to localize a complicated concept, a whole mental faculty of a

that we believe that split consciousness does indeed occur but under ordinary conditions is not influential because the dominant (left) hemisphere does not appear to be that concerned with or aware of the activity of the minor hemisphere (Galin, 1974). One exception to this is most probably the case of psychoanalyzed persons, whose left hemispheres learn how not to ignore right-hemisphere signals (also see Note 12).

> psychic element? Is it justified to immerse a nerve fiber, which over the whole length of its course has been only a physiological structure subject to physiological modifications, with its end in the psyche and to furnish this end with an idea or memory? [pp. 54-55].

Freud (1891b) suggests that "a mental phenomenon corresponds to each part of the chain [of events in the nervous system], or to several parts. The psychic is, therefore, a process parallel to the physiological, a 'dependent concomitant' " (p. 55). Or, put differently slightly later in his text, he states that "the localization of the physiological correlates for perception and association is, therefore, identical." (p. 57). Speech is seen as something "of a complexity beyond comprehension," yet this complexity is handled within the limits of the then current neuropsychiatric knowledge on the basis of functional considerations.[7] It is this same functional viewpoint of localization that we are about to apply in our discussion of brain asymmetry. Later we will return to the subject of aphasia to give specific clinical examples of hemispheric-disconnection states with neuropsychiatric implications.[8] We speculate that the unconscious, affectively important aspect of mind is shared by both hemispheres but possibly with the greater contribution coming from the right (nondominant) hemisphere (Wexler, 1980), whereas certain conscious recognition, or verbalizable experience, of external or internal events appears to have a greater contribution from the dominant (usually the left) hemisphere. That is, it seems to us that the best evidence favors the assumption that, although conscious mental activity is the product of the system of two hemispheres collaborating together with the rest of the nervous system, in this collaboration there is reason to believe that *crucial aspects of the underlying mental processes* (only a small part of which are conscious) *are partitioned between the hemispheres,* with the dominant

---

[7]Freud's work on aphasia not only played a role in revising neurological concepts of localization, however; it also played a decisive role in the establishment of psychoanalysis itself (see Binswanger in Freud, 1891b). In fact, neurological concepts of disconnection lead to the topographic and structural models of the mind, in which certain affects are seen as becoming functionally disconnected from ideas (e.g., obsession and isolation of affects) or in which both ideas and their associated affects are understood to be disconnected from consciousness (e.g., repression and disavowal). Moreover, in *On Aphasia* Freud anticipated many later developments in psychoanalysis: paraphasia ("the psychopathology of everyday life"), the speech apparatus (the "mental apparatus"), projection, representation, overdetermination, regression, *Besetzung* (translated later as cathexis), and others.

[8]Those interested in a review of the general advances in aphasia research since Freud, of which there have obviously been many, should consult Brown (1969) and later chapters of this book.

(usually left) hemisphere following the classical secondary-process mode and the nondominant (usually right) hemisphere following the classical psychoanalytic primary-process mode of function. Freud's (1985) speculations along these lines in the "Project" perfectly anticipate the core cerebral asymmetries under discussion (see especially McLaughlin, 1978).

We might ask how this mental processing is partitioned, that is, how the integration of the two cerebral hemispheres actually occurs. The best discussion of this speculative subject seems to be Galin's (1974). He considers essentially four possibilities:

1. The hemispheres operate in alternation, "taking turns, depending upon the situation demands. When one hemisphere is 'on' it inhibits the other" (p. 575).
2. The dominant hemisphere "makes use of one or more of the subsystems of the other hemisphere, inhibiting the rest. The inhibition thus may be only partial, suppressing enough of the subordinate hemisphere as to render it incapable of sustaining its own plan of action" (p. 575).
3. "One hemisphere dominates overt behavior, but can only disconnect rather than totally inhibit (disrupt) the other hemisphere, which remains independently conscious" (p. 575).
4. The two hemispheres are fully active and with each other.

The fourth possibility Galin associates with "creativity, man's highest functioning" (p. 575). We would like to add the possibility that this improved hemispheric integration may also be associated with successful analyses. The first three possibilities are considered by him to be characteristic of the "reciprocal inhibition" of sensorimotor systems "around which the whole brain is built" (p. 575).

Galin further believes that two factors determine which hemisphere actually dominates. Reviewing a number of experiments in split-brain subjects and in primates, he considers the determining factors to be the speed of problem solving and the motivation of the competing hemispheres. By motivation he is referring to the ability to earn reinforcement. Thus, the left hemisphere is alleged to obtain ascendancy over the right hemisphere at age two or three years because of its success and speed at using expressive language to manipulate the environment.

Interestingly, Ferenczi (1926) felt that the left side of the body (i.e., the right hemisphere) seemed more in touch with the unconscious, an observation that fits current neuropsychiatric knowledge. In this regard it has been observed by Sackheim, Gur, and Saucy (1978) that emotions are expressed more intensely in the left side of the face (i.e., under the control of the right hemisphere). There is reason to believe that the so-called artificial tripartite categories of brain activity, which go back at least as far as Aristotle—thinking, feeling, and behaving—may themselves merely express the bicamerality of the mind/brain, with thinking and analytic processing representing the left hemisphere and emotional coloring being provided chiefly by the right hemisphere, or perhaps conjointly with the limbic system and its commissures. The final behavioral and experiential outcome, of course, depends on optimally combined activity of all the brain's systems for information processing[9] and affect regulation, whether spinal, brain stem, midbrain, thalamic, hypothalamic, limbic, or bicameral/cortical.

In other aspects of language function, the dominant left hemisphere is responsible for speech formulation (in its expressive or receptive phase), but the nondominant right hemisphere seems to provide the skill for optimal prosodic delivery and interpretation (comprehension) of speech. Thus, although superb, the dominant hemisphere's delivery may nevertheless fall short of providing emotional impact without the prosodic contribution of the nondominant hemisphere. (This is an example of the complexity of the interhemispheric collaboration and idiosyncracy referred to earlier). Gestural characteristics and body language are also the function of the right (nondominant) hemisphere, and this adds still another element inproving optimal communication skills. That each hemisphere seems to have its own communicative modes also opens the door for conflicting signals that might be sent simultaneously, resulting in confusion for the recipient of such mixed messages (Galin, 1974).

ADDITIONAL DIMENSIONS

The reader will appreciate that we are oversimplifying, for the sake of presentation purposes, the extremely complex subject of cerebral integration. Because of our selection out of the whole fabric of nervous system activity those dimensions that follow lateral (left–right) asymmetry, the impression is artificially created that this is the only, or even the major, way to understand cerebral phenomena. Clearly, such a view overlooks the triple-brain mode of MacLean (1960), the hierarchical models of Jackson (1958) and of Gedo and Goldberg (1973), and many other integrative views. For example, we have left out completely from

[9]Work with commissurotomy (split-brain) patients further shows that short-term memory is invariably impaired. This finding lead to the conclusion that the connection between the hemispheres is important not only for information transfer but for the fixation of memory itself (Zaidel and Sperry, 1974).

the present discussion (but hope to redress this weakness in the future) consideration of neurohumoral, neurotransmitter, and other chemical systems of organization.

THE ROLE OF METAPHOR

In Chapter 1, on metaphor, some basic mechanisms of effective psychoanalytic interpretation were described. Such descriptions were extensions of original insights of Strachey (1934) and attempted to reconcile this older work with modern psychoanalytic theory, as well as with what is known in psycholinguistics, Piagetian psychology, and the neurophysiological work of Lassen, Ingvar, and Skinhøj (1978). It was shown that the analyst's affectivity has general as well as specific effects and that metaphors in the analyst's transference interpretations have precise specificity in arousing a patient to a state that allows for synthetic activity. Metaphors serve as "bridges" in a number of ways. First they allow for the linking up of the two hemispheres. This results from the fact that metaphors constitute an ambiguous stimulus object that can arouse activity in each hemisphere by appealing to the left hemisphere linguistically and to the right hemisphere by nonlinguistic means. Metaphors thus come close to being simultaneous translations of themselves; they provide functional neuroanatomical bridges or pathways that will never be seen under any microscope but that help the brain hemispheres collaborate in their activities.

As we shall discuss, we are coming to the conclusion more and more that the connecting of the two hemispheres overcomes repression and/or disavowal. It is also our position that psychoanalysis, especially, sets up the conditions for this connection to occur through the use of metaphors and through many other means as well. What will be decisive in the coming years will be to identify most carefully which brain diseases, including psychiatric illness as a critical subgroup, represent hypoconnection states that are remediable by psychoanalytic means and which do not.[10] In our view, this job will require neurologists *and* psychoanalysts to work together without biases against each other's viewpoints. It is possible that different commissures are impaired to different degrees in different conditions, and such techniques as psychoanalytic studies, either independently or in tandem with such techniques as nuclear magnetic resonance imaging (Pykett, 1982), offer the prospect of eventually

[10]Please note that we are not assuming an either/or relationship here. As postulated in Footnote 5, there is experimental evidence to suggest the theory that in addition to functional changes during analysis (learning), there may also be structural changes in the brain.

answering such questions more definitively in the decades ahead. In this regard, we may be closer than man has ever been to understanding remembering and forgetting in terms of fundamental psychological and neuroanatomically related mechanisms (see Note 9).

LANGUAGE, COMMUNICATION, AND THE BICAMERAL MIND

Language and communication are different; the former is just one example of the latter, larger domain. Thus, although the left hemisphere is most often associated with language, it is still possible to imagine that each hemisphere might communicate to us or be communicated with by independent means.[11] It is generally conceded that the left hemisphere primarily uses formal language and appeals to logic in communication. (We assume that slips of the tongue or in writing would thus tend to show left-brain thinking or other affective influences perhaps processed in the right hemisphere and then conveyed to the left hemisphere. It is interesting to contemplate whether the left hemisphere, through such reverberating means, effectively tones down right-hemisphere affect. Since in our opinion the two hemispheres are always connected functionally to some extent, we are dealing with leading zones in a bihemispheral relationship.) In contrast, the right brain communicates by nonverbal, nonlogic-based means, for example, by facial expression, by receptive prosody capabilities, and by prosopagnosis. In this regard, it has been speculated that interpersonal conflicts can be created by mixed communications that are interpreted (decoded) differentially by the two hemispheres of the person receiving the communication, as in "your eyes say yes, but your words say no, no, no!" (see Galin, 1974).

TRANSFERENCE AND BICAMERALITY

It would be intriguing if transferences themselves might be sometimes reflective of the integrated activity of the entire brain but at other times of only a part of it, say of only one or the other cerebral hemisphere. Since the psychoanalytic situation involves a time regression and since the myelinization process results in one period (from birth to age three and a half) during which the cerebral hemispheres are presumably maximally iso-

[11]One clue as to which hemisphere is dominant at a given time may be the turning of the eyes and/or the head (Kinsbourne, 1972; Galin and Ornstein, 1974). In solving verbal problems right-handed people are alleged to show an initial turning of the head and/or eyes to the right, and in solving visual/spatial problems, to the left. The reverse is true of left-handed subjects.

lated from each other and another period (from age three and a half to adolescence) of only relative hypoconnection, it seems logical to conclude that at times transferences might represent one or the other hemisphere in more or less "pure culture."[12] If during the preoedipal period the hemispheres are functionally hypoconnected, the probability exists that crucial affects are being separated from their age-appropriate ideas or meanings.

Now, the question must be raised whether these hypoconnection states result in danger or an advantage for the organization of the personality. Although hypoconnection may be seen as constituting a danger, this, we would contend, is not necessarily so. For example, it is possible that during certain periods, hypoconnection may function as a stimulus barrier, protecting the organism from certain affects, at least until a different psychophysiological mechanism for modulating the affects involved is developed. After the appearance of an affect-regulating psychic structure, the disadvantages of disconnection of the cerebral hemispheres might begin to outweigh the advantages. From this vantage point it may be that disconnection itself functions as a defense, or adaptive neurological mechanism, especially during some of childhood (Grotstein, 1981). For each person the question of what is optimal might therefore depend on the balance of two parallel developmental trends: the growing need and capacity for coordination and integration of hemispheric activity, and the need and capacity to develop coping mechanisms to avoid the danger of being chronically overstimulated. It seems safe to conclude that the more rapidly and reliably the individual hemispheres develop tension-regulating ability, the sooner their connection will form a more optimal system. The presence of another brain (either parent's or therapist's) could be hypothesized as helping the child/patient avoid depression or anxiety by taking actions that appropriately favor or inhibit interhemispheric connection. Thus, in some cases *one hemisphere might not know what is in the other*. This point cannot be overemphasized, since it is the basis of a functional definition of disavowal and repression as interhemispheric disconnection states.

At this point it may help, however, for us to give brief examples of what might be meant clinically by the preceding statement. Certain kinds of transferences might relate to the dominance of one hemisphere over another. The first example is one of Sperry's as described in an article by Galin (1974); the patient has had her corpus callosum sectioned:

[12]The psychoanalytic clinical phenomenon of fragmentation may at times represent a true split-consciousness in which there are differing realities in awareness, some from one cerebral hemisphere and some from the other; and the situation is complicated still further by different realities even in one hemisphere based on the self experience from different developmental periods.

> One film segment shows a female patient being tested with a tachistoscope. . . . In the series of neutral geometrical figures being presented at random to the right and left fields, a nude pin-up was included and flashed to the right (nonverbal) hemisphere. The girl blushes and giggles. Sperry asks, "What did you see?" She answers, "Nothing, just a flash of light," and giggles again, covering her mouth with her hand. "Why are you laughing then?" asks Sperry, and she laughs again and says, "Oh, Dr. Sperry, you have some machine!" The episode is very suggestive; if one did not know her neurosurgical history, one might see this as a clear example of perceptual defense and think that she was "repressing" the perception of the conflictual sexual material. . .[p. 573].

A second example is a patient in analysis who began to talk in a detached way. Her analyst's attention wandered to the patient's left hand, which was shifting periodically into a "gun" gesture entirely out of her awareness. When this gesture was called to the patient's attention, she responded by remembering that she had read in graduate school about right-left brain studies, including Ferenczi's observation that the left side of the body (i.e., the right hemisphere) appears more in touch with the unconscious. (The patient did not know of the analyst's interest in the subject of left-right brain studies.) She simultaneously became aware of and experienced significant anger toward the analyst, which during the session she gradually was able to relate with increasing certainty to the analyst's upcoming vacation. Finally, at the end of the hour, she began to recall some specific memories of being taken advantage of sexually by certain parental substitutes in childhood.

Let us now explain more exactly about disavowal and repression. Writing on repression, Galin (1974) proposes "that in normal, intact people mental events in the right hemisphere can become disconnected functionally from the left hemisphere (by inhibition of neuronal transmission across the cerebral commissures), and can continue a life of their own. This hypothesis suggests a neurophysiological mechanism for at least some instances of repression and an anatomical locus for the unconscious mental contents" (p. 581).

Writing on the subject of disavowal from a strictly psychoanalytic perspective, Basch (1983) describes episodic and semantic memory processes (the former is based on personal memory, built on affectively charged feedback cycles, and the latter, roughly speaking, coincides with general knowledge or deductive thinking), and he relates these psychological categories to the characteristic qualities of the two cerebral hemispheres. According to Basch:

> The consciousness of self that we call "I" requires that the right brain self-experience, the episodic memory, be translated into verbal or other forms of discursive language. *In repression it is the path from episodic to semantic memory, from right to left [brain], that is blocked.* The self-experience can no longer be articulated and, therefore, cannot be either thought about or made conscious. *Disavowal works to block communication in the opposite direction, from left to right.* That is, what is apprehended on the semantic, left-brain level in terms of language and logical categories cannot be translated into right-brain, episodic, self-experience. As Freud postulated, in both repression and disavowal affect is "eliminated" [p. 151; italics added].

In the patient mentioned earlier the woman who was gesturing angrily with her left hand because her right hemisphere had some upsetting experience that could not be articulated, certain memories and their associated affect were anatomically isolated (repressed) until she and her analyst observed her gun gesture as a communication from the right hemisphere and began to consider its significance. The gun gesture itself and the affect connected with it had originally been disavowed. The interpretive sequence was first to interfere with the disavowal; this led to the uncovering of repressed material (the forgotten upcoming vacation of the analyst and the forgotten sexual experience in childhood).

PSYCHOLOGICAL CONFLICT AND BICAMERALITY

It follows from the foregoing discussion that conflicts can potentially involve discrepancies either within or between the cerebral hemispheres; thus, the classical psychoanalytic models (topographic, structural) are entirely compatible with the latest neurological knowledge. The complexities and variance of hemispheral hypoconnections might also explain why it is that even in successful psychoanalytic treatments conflicts do not appear ever to phase out completely (Pfeffer, 1963; Schlessinger and Robbins, 1975). At a basic level, behavior and inner experience are always the summation (or compromise) of trends in two cerebral hemispheres. The two hemispheres can, anthropomorphically speaking, learn to collaborate with each other more effectively and cycle their "disagreements" or styles of function more rapidly, but the dichotomous stamp remains on all the products of their joint activity.

Since transferences may represent such a joint product of both hemispheres, it behooves us to ask whether one hemisphere may be dominant in talking to us and which hemisphere we predominantly wish to reach. This is another reason why it may be crucial to allow, when it is appropriate, the use of metaphors in our transference interpretations: these will appeal to both hemispheres simultaneously, thus providing a bridge between semantic and episodic memory systems, between dominant and nondominant hemispheres. The cohesiveness that results may be the direct outgrowth of more optimal interhemispheric information transfer, integration, and collaboration, as well as of critical changes in the "software" within the hemispheres individually or in the overall system they form. We may be accused of reductionism in saying so, but the feeling we get is that by addressing both hemispheres simultaneously we are talking to the whole person.

APHASIA AND BICAMERALITY

Let us return again to the subject of aphasia, which we examined first in regard to Freud's book *On Aphasia,* one of his earliest works in neuropsychiatry and one that seems to us of critical importance in understanding the subject of the bicameral brain. First, we will give some clinical neurological examples, which we believe will further show the utility of our working assumptions regarding disavowal and repression as psychological clinical phenomena that probably relates at the neurophysiological level to functional dysconnection states. Then, we will attempt to tie together our discussion of brain bicamerality by including some final speculations on the origin of bicamerality, by indicating the place for an integrating mechanism and by pointing out what we think this might be (see Frick, 1982). Last, we shall summarize this complex subject, ending with the hope that our efforts, admittedly imperfect, will be seen not as bad psychiatry or bad neurology, but rather as a reflection of the imperfect state of our neuroscience today.

Specifically, we wish to consider the special subject of aphasia in multilingual people, drawing primarily on the monumental study by Minkowski (1963), who drew particularly from A. Pitres (a pupil of Charcot). Pitres described (in 1895) how polyglot persons recover their languages in a special pattern: they understand, then learn to speak again, first the language that was most current at the onset of the aphasia. Pitres felt that this systematic pattern of loss and return of language—which has become known as Pitres's principle—constituted evidence against the existence of separate centers for each language. Pitres further noted, in his seven original cases, that restitution of the linguistic function occurs only when the speech centers are not irreversibly damaged; that is, he was describing a functional disturbance that he felt was secondary to the temporary inactivity of cortical centers.

According to Minkowski (1963), Pitres's conclusion retains its full significance; however, there have always remained so-called exceptional cases that do not really fit the rule under certain "neuro- and/or psychobiological conditions" (p. 120), cases in which one language is favored (retained) over another (which is slow to return).

In our opinion, these exceptional cases appear to be clarified if one takes into account repression and/or disavowal and if one includes in the appreciation of these psychoanalytic mechanisms their possible physiological correlates, as we have hypothesized them earlier, that is, functional cerebral disconnection states that isolate the right from the left hemisphere (repression) or the left from the right hemisphere (disavowal).

Minkowski (1963) wrote of the exceptional case of a 44-year-old railway conductor, a Swiss-German who spoke, along with his native and literary German, a little French, which he had learned at school and from his father, who knew French well. From the ages of 19 until 25 he lived in France, where he fell deeply in love with a French girl, with whom he had an affair that extended over two very happy years. The patient stated that this was his first and greatest love.

On returning to Switzerland where he spoke his native German, the patient married, had three children, and then at age 44 suffered a stroke with a right hemiplegia, Broca type. Contrary to Pitres's principle, the first language to return was French, and he initially conversed in and understood only French. Then, his Swiss-German and literary German gradually returned over a period of months, always with some regression of his French language proficiency. Since we believe the details are important, we will quote from a section of Minkowski's (1963) description:

> The evolution of the aphasia in this case has been particularly remarkable. Comprehension of all languages appeared more or less restored already one to two days after the apoplectic attack, whereas disturbance of expressive speech proved to be very severe and peculiar. . . . [When he spoke he spoke French] his wife did not understand him and his children with their poor school-French acted as translators between their parents. Three weeks after the apoplexy he spoke a rather correct French and only then he began to speak also German, but only the literary language. As to the Swiss dialect [of German], it first appeared in patches, only four months after the fit, and even then progressed slowly. At the same time he began to read again, principally German papers. Six months after the stroke he spoke fluently a relatively correct French with some disturbances of motor aphasia, his German was still less advanced than French; the Swiss dialect [of German] was still spoken only in fragments. . . .

> A few weeks [after this six-month phase], during the Christmas holidays which the patient spent at home, *in an intimate family atmosphere,* he began to his own surprise to speak his native Swiss dialect almost fluently at the same time the French gradually regressed and [he] sadly regretted it [p. 130; italics added].

Another exceptional case (Minkowski 1963) is a 24-year-old Jewish male who was reared in Germany until the age of 20. He attended high school and later taught in some kind of seminary school. At the age of 22 he emigrated to Palestine, but he continued to speak primarily German with his friends and family while gradually acquiring a knowledge of written and spoken Hebrew. After a traumatic injury to his left temple, he remained unconscious for five days. Upon awakening he uttered in German, "Dear only God, thank thee eternal" and several other automatic utterances, after which he spoke only in Hebrew. His Hebrew, though initially hesitant, gradually improved. In addition to a sensory aphasia he had a right hemianopsia. Four months after his injury, the patient was writing and speaking Hebrew fluently. In contrast, his German did not really return with fluency, in spite of its being his predominant language at the time of his injury.

We agree with Minkowski (1963) that the first case probably represents "hemorrhage on the left side in the anterior segment, the genu of the internal capsule, and the corresponding part of the corona radiata" (p. 13), as well as involvement of left-sided frontal areas of speech (possibly Broca's convolution, the medial surface of the left hemisphere in front of the paracentral lobule, and possibly some other lesion of the "speech apparatus," if one would use Freud's terminology). But in the spirit of Freud's pioneering work (on localization) in neurology and in psychiatry, we cannot emphasize sufficiently the role of unconscious mechanisms (especially of repression) in this case. And, although Minkowski in his discussion mentions in passing how this case "can obviously not be explained by purely linguistic factors" (p. 131) and suggests instead "the active cooperation of particular psychic agents in its determination," we believe Minkowski did not take full advantage of his clinical data and their neuropsychiatric implications.

First of all, the patient's greatest (heterosexual) passion in his life was with the Frenchwoman. Apparently, the love for her needed to be repressed, and presumably along with this repression there was a need to disavow the emotional significance of the French language, the use of which would have pointed to the presence of this repression. Repression was apparently successful until his injury. The left-sided stroke would have then facilitated the emergence of French for a number of reasons: First, if repression represents, as we believe, a sequestration of some right-hemisphere affectivity away from the awareness of the left hemisphere, then, with the left-hemisphere (comprehensive/expressive) language function eliminated temporarily by the stroke, there would be no longer any danger of the right-hemisphere affect being comprehended semantically by the left hemisphere. [13]Second, with the language capacity of the left hemisphere temporarily lost, the only language function (available or employable) is that of the right, presumably along with its high emotional loading for French; that is, from the point of view of the surviving right hemisphere, French is more important than German because the French lover was more loved than the patient's German-speaking spouse. Third is the possibility that the right hemisphere's affect is capable of being fully appreciated by the right hemisphere itself but that this comprehensive function is slower to recover than the expressive function.

There are numerous other possibilities; for example, as the patient recovered he may also have needed to repress his relative lack of affection for this German wife. But we have mentioned the major ones, with the exception of one additional factor that is significant enough to speculate about. Minkowski noted that for this first patient French was actually originally connected with his father. This pattern of recovery suggests to us the hypothetical possibility that at a still deeper level within the unconscious (represented here, at least in part, by the right hemisphere) there was a repression of homosexual feelings toward the father. This valuation of the French lover would have provided an additional impetus, in our opinion, for the reemergence of French from the isolated right hemisphere during the recovery phase of the illness, as well as for the reemergence of German (along with the simultaneous loss of French) later, presumably when the left hemisphere recovered. These language changes would reflect not only the recovery of neurological function as mediated by the language-dominant left hemisphere but the reestablishment of repression and disavowal (of the repression)[14] in the form of

---

[13]This might also be considered a release phenomenon in which the damaged left hemisphere is seen as failing in its ordinary function of suppressing the right-hemisphere (repressed) affect. However, a more likely possibility is a mere disconnection of the two hemispheres that are ordinarily connected. We see no reason for believing a priori that one hemisphere always suppresses the activity of the other. (See the following paragraphs, however, for a fuller enumeration of the complex possibilities.)

[14]From our viewpoint, disavowal would, as has been clinically understood in psychoanalysis, invariably occur along with repression—almost never would these mechanisms occur in isolation. From the view of the clinical situation, this means that if there is a repression we also need to avoid awareness of (i.e., to disavow) anything that points to the existence of the repression. From a neu-

functional disconnections of the two cerebral hemispheres, which would keep the patient from becoming aware of unacceptable heterosexual and/or homosexual impulses or feelings.

At this point, let us move on to Minkowski's second case of exception to Pitres's principle, the one involving a 24-year-old Jewish man (born in 1917) who came to Germany "as a nursling." This man's history should be expanded to note that he learned Hebrew as a child "from prayers and the Bible" although it was only later, in Israel, that he began to use Hebrew in any significant way. The bullet that struck his left temple in 1939 tore the dura mater and the underlying brain tissue as well.

Following Pitres's principle, this patient should have first understood and then spoken German during his recovery. In fact, his first comment was in German but it was brief, and only later did his German return to a significant degree. Instead, the priority of Hebrew over German lasted eight months posttrauma. Why? The record does not state whether the nurses and doctors communicated with the patient primarily in Hebrew although this is likely the case and doubtless contributed its own effect, for example, by helping to fill roles as (Jewish) transference figures. According to Minkowski (1963), the German "repertory" was "inhibited and repressed by his Hebrew," and the Hebrew was "preferred" for social and affectional reasons "in face of the impossibility of a simultaneous restoration of two different languages in a damaged common centre [sic] of speech" (p. 141). In our opinion, Minkowski correctly notes the factor of the then flagrant anti-Semitism in Germany and the patient's Jewishness, which common sense tells us clearly played a role; however, Minkowski is unable to specify what this role was. He ends up making more and more complex a priori assumptions, such as that it is impossible to have a simultaneous restoration of two languages.

It would seem easier to say that though Pitres's principle generally obtains, one must take into account the state of repressions (and/or disavowal); that is, the role of the nondominant hemisphere is crucial. This consideration explains that this undamaged hemisphere expressed itself, in the case of this Jewish man, by the use of Hebrew rather than

rophysiological point of view, if one has a need to keep the activity of one cerebral hemisphere from the other, then the flow of information would probably need to be blocked in both directions at the same time or there would always be the possibility of information being inferred by the observation of the feedback to other outgoing information. This pattern of control is the same as the example of the Russian government, which protects itself by keeping undesirable visitors out, but which also controls the outflow of its own people, who bring unacceptable information back with them from travels abroad.

German. We would argue, however, that his disdain for German was really not repressed but disavowed. (Of course, because of the dearth of psychologically relevant data and details on the patient's personality, it is impossible to offer anything approaching a comprehensive psychodynamic formulation. But a number of mechanistic-sounding suggestions will be attempted.)

From our point of view, his language difficulty involves a disconnection syndrome in the left-to-right direction. As we stated earlier, quoting Basch (1983), in disavowal "what is apprehended on the semantic left-brain level in terms of language and logical categories cannot be translated into right-brain, episodic self-experience" (p. 99), and thus there is an interference in the fundamental depth of the experience of the self.

So we propose that this Jewish man could not speak German for a number of reasons. First, he is disavowing the significance of German to himself. Second, he needs to disavow German because there is some associative chain that relates to the German language that is itself part of the repressed material (possibly the Holocaust experience touched on this).[15] If he were to become aware of any understanding of German, he would presumably also become severely anxious by virtue of contact with repressed affect, so instead he has to rely only on Hebrew, with its entirely different and presumably more pleasant set of associations (the flight to Palestine from Germany etc.). Possibly, the initial statement in German[16] was sufficient by itself to arouse his need to strengthen the repression of German (to block out further information flow in the left-to-right direction). In contrast to the previous case, there is also always a possibility that the left hemisphere in this second case is either better able to understand, that is, less damaged (which would lead to a greater need to block the left-right flow), or that the second case is different from the first in some other fundamental mechanism as well (for example, this second case was labeled by Minkowski a "sensory aphasia" whereas the first case was called a "Broca's aphasia"). Since this chapter is not fundamentally on aphasia but rather on relating the significance of right–left brain studies to psychoanalysis, we will end our discussion on aphasia at this point.

[15]The case history does not state that this man fled Germany to avoid the Holocaust, but we think the timing of his move makes this a likely conclusion.

[16]It has been speculated that the right hemisphere does not have the phonemic analyzing ability of the left (Levy and Trevarthen, 1977; Sidtis et al., 1981). Disconnected from the left, it thus can usually initiate no more than a single syllable or a "habitual, multisyllabic phrase," as is true in this case.

THE EVOLUTIONARY SIGNIFICANCE OF BICAMERALITY

We would like to return briefly to the evolutionary, or adaptive, significance of bicamerality. As Levy (1974) describes so eloquently, the archaeological evidence unequivocally shows that in Africa in the Pliocene period Australopithecus killed baboons for food by crushing their skulls with "weapons" held in his right hand; that is, our ancestors were already manifesting cerebral asymmetry (by right-handedness) two million years ago! Levy (1974) feels, and we agree, that cerebral lateralization resulted in an evolutionary advantage for two major reasons: (1) a population of mixed phenotypes would be able to specialize, with a predominant group of mostly hunters (with dominant right hemispheres—good for depth perception, visual memory, gestalt closure, and directional discrimination of movement) and a smaller number of planners, "with functionally symmetric hemispheres," (p. 174); (2) in addition, there is almost a doubling of the overall cognitive capacity of the individual human brain that results from the division of labor between the hemispheres (Levy, 1977). We also wish to add to Levy's remarks our belief that symbolization and meaning become possible only after mental words, pictures, and their accompanying affects can be tied together, as they are uniquely in our brain's bicameral structure (see Galaburda et al. 1978). Unfortunately for humans, that which has adaptive advantage also opens the door for intrapsychic "conflict," as discussed in this chapter under the rubric of repression and disavowal. It should be obvious from our discussion of the language disturbance called aphasia that each cerebral hemisphere makes its own unique and critical contribution to our potential to communicate with others and ourselves and especially that the two hemispheres must collaborate, or the overall communication system fails in its adaptive tasks. It was Freud who first pointed out the special clinical significance of repression and disavowal as mental (defense) mechanisms; we have tried to show here that these mechanisms are probably over two million years old and flow from the asymmetry of our brains (they are linked to directional disconnection states of the two cerebral hemispheres). We believe that this correlation aids not only the understanding of the complex subject of aphasia, but also enhances our comprehension of mental mechanisms in general.

"Mind," according to Lashley (1951), is "a complex organization held together by interaction of processes . . . it has no distinguishing feature other than its organization" (p. 136). We have tried in this paper to clarify some aspects of this unbelievably complex organization of mind. Clearly, one property of the nervous system is its capacity to experience and manifest intrapsychic conflict. As we have suggested, conflict can also mean the relative disconnection of the two (asymmetrical) cerebral hemispheres.

SUMMARY

We have reviewed some neurological knowledge of the two cerebral hemispheres and integrated it with psychoanalytic theory and practice, raising a number of theoretical questions and suggesting some answers. We have related the onset and pattern of myelinization of the interhemispheral connecting tracts to the watershed period of emotional development, the oedipal phase. We have related analyzability to brain plasticity as a possible decisive factor.[17] We next discussed hypoconnection states and the data of hemispheric idiosyncracies, pointing out the dangers of oversimplifying complex system-relationships and of using outmoded ideas of localization of brain function. Generally, we found that the topographic and structural psychoanalytic models were entirely compatible with the latest neurological understanding of the bicameral brain. Starting with Freud's work on aphasia (which was a precursor to these crucial psychoanalytic models) and his ideas on primary and secondary process (which anticipated much of the work of split-brain studies!), we went on to discuss the role of a specific interpretive element (with linguistic and nonlinguistic properties), namely, metaphor. Following up on the bridging role of metaphor in the analyst's transference interpretations, as discussed in chapter 1, we described how this bridging of metaphor (and by other means psychoanalytically) applies specifically to states of disordered connection of the two cerebral hemispheres. In so doing, we developed neuropsychiatric hypotheses regarding the subjects of disavowal and repression. We pointed out how the same conclusion has been reached by independent investigators (Galin, 1974; Basch, 1983), who each started from a theoretical position different from ours, lending support to our ideas about repression and disavowal. Finally, we considered clinical examples of communications from and with the bicameral mind, taking up as well exceptions to Pitres's principle. We believe these so-called exceptions *can* be explained by the use of psychoanalytic perspectives, especially by considering repression and/or disavowal mechanisms as they are described and defined in this chapter.

We feel that what proves crucial at times in the psychoanalytic situation is the possibility of the analyst's serving as a functional link

[17]By plasticity we are referring to the capacity to learn and grow emotionally, as reflected by the modifiability of the anatomy (cytoarchitectonic details) or altered functioning of the modular systems of the cortex (Szentagothai, 1975). The subject is too important and complex for further discussion here.

between the analysand's two cerebral hemispheres. We believe there are good reasons to expect that the addition of metaphors in the analyst's (transference) interpretations in this role as a catalyst (of a bidirectional process) will enhance the analytic process by serving as such a bridge. Along the way we implied that the meaning of conflict can be expanded to include the situation of suboptimal interhemispheric connectedness (although this is clearly not the explanation for all conflict) and thus might explain the well-known observation that even after successful analyses, psychic conflict is not entirely obliterated.[18] There are always two cerebral hemispheres; even if they can learn to cooperate with one another, they still have their own stylistic differences or idiosyncracies. Finally, we have made reference to integrating articles by Galin (1974) and Frick (1982) that indicate some ways in which interhemispheric regulation might be accomplished. (See Galin for further details regarding "reciprocal inhibition," cerebral dominance by means of speed, and integration through various levels of the vestibulocerebellar system.)[19] We hope that our attempts at imbricating neurological with psychoanalytic concepts are useful in the expansion of both fields, a task we claim no less the property of one field than the other.

---

[18]Of course, other factors or viewpoints may be seen as playing a role in the preservation of premorbid difficulties—for example, fixations, the role of trauma, the need for autonomy, arrests in development that have not been influenced by the psychoanalytic process, etc.

[19]We have left aside a complete discussion of the important question of why cerebral disconnection is perpetuated once it occurs, i.e., after myelinization of the intercerebral connecting tracts. We suggest that the answer hinges on understanding how the cerebral hemispheres integrate their activities, something that is unknown but that we have speculated about, quoting Galin (1974) in the section on Hemispheric Stereotypes and the Case of Aphasia.

3

Brain Plasticity, Learning, and Psychoanalysis

Some Mechanisms of Integration and Coordination Within the Central Nervous System

with D. M. Vuckovich

PRÉCIS

The previous chapter ends where this begins, with the question of how the cerebral hemispheres are coordinated (bridged) in their activities. A second major question about the nature of learning is also pursued. These subjects are considered important because therapeutic process appears to unlock our potential for learning, and learning seems to involve some process in which the various learning subsystems of the brain are able to exchange data. Whether one is employing a theory of three brain subdivisions (as MacLean's Triune Brain Theory does), two major subdivisions (as in a hemispheric model), or any number of major subsystems, one must still solve this problem of coordination between parts. The neuroscience center of this chapter is the work of the Japanese physiologist Itoh on the cerebellum, that part of the brain which oversees important aspects of "neural control." Itoh's work makes clear how cerebellar-aided decisions may be made that match problems a person faces with brain structure suited for adaptive decision making (problem solving). Simply stated, there is evidence that the cerebellum plays a major role in learning. There is also evidence of the construction of a cerebellar-based model of self-in-the-world the manipulation of which allows thinking without touching (or mouthing) and the structure of which provides a core sense of self. To appreciate more fully what learning is, a significant part of the chapter is devoted to the subject of brain plasticity, a word that describes the ability of the brain to capture experience. This subject is possibly too complex for the short presentation in this chapter, but

I believe a review of the details gives some indication of the brain's hierarchical organization (a subject of several later chapters). One last note: To check my theory out, I personally consulted with Itoh, reviewed this chapter with him, and assured myself that I have properly understood Itoh's work and its implications regarding learning. In addition, some recent work of Itoh seems to focus more on my interest and perspective, namely, the general issue of the relationship between cerebellar neural control and psychological issues of adaptation. Clearly, the communication between psychoanalysis and neuroscience assists both fields.

BRAIN PLASTICITY

"The nervous system of mammals retains throughout the animal's life span the ability to modify the number, nature, and level of activity of its synapses" (Cotman and Nieto-Sampedro, 1984, p.1287). Reviewing plasticity, Hagan (1984) comments on Purves's view of the human brain as an ecosystem, not as "hard-wired" like that of lower animals. "The connections between mammalian nerve cells are established with a great deal more flexibility, and [this situation exists] over a surprisingly long period of time" (p.4). Evidently, the synaptic deck is capable of being continuously reshuffled, and each neuron, rather than getting a pat hand, participates in a kind of survival of the fittest, in which many mechanisms create a flexible and dynamically changing pattern of connectivity. Some of these mechanisms are chemical, such as the trophic substance Nerve Growth Factor (NGF), discovered by Levi-Montalcini and Victor Hamburger. Other factors include the response to "learning", as illustrated in the work of Merzenich and associates (1984) on changing somatosensory maps and the work of Nottebohm (Nottebohm, 1985; DeVoogd and Nottebohm, 1981; Kolata, 1984), which has now specified some of the incredible learning changes in the frontal cortex of adult birds. Together these studies illustrate that *the human brain is a self-organizing and plastic organ that continues to change and adapt throughout life,* not a static machine with unchangeable "hardware" (Racine and Zaide, 1976; Moore, 1979). Moreover, although some factors do limit growth within the adult central nervous system, there is no question that central neurons retain "regenerativity" (Black, 1984, p.25).

In considering "plasticity" we shall adhere to a slightly modified version of the general schema of Bloom (1985), who summarizes the kinds of complex developmental and postinjury plastic changes as follows: (1) effects of neurochemically defined signal transmission affecting transmitter molecules and/or receptors; (2) both excitatory and inhibitory nervous system effects on aggregates, such as the reticular activating system, with associated impedance changes in various target cells (e.g., the locus ceruleus, involving cerebellar, hippocampal, and neocortical changes in norepinephrine); (3) effects on individual neurons, including altering such intracellular elements as cyclic AMP, protein kinase, macromolecular substrates, ion channels for calcium in the pre-and post-synaptic membrane, and so forth; (4) the increase in neuronic complexity postnatally that is part of normal development; (5) the construction of short-and long-term memory circuits during development, after injury, or in relation to some novel 'learning situations'; (6) new growth of central neurons secondary to trophic influences, such as nerve-growth factors; (7) compensation by means of activity in areas outside of an area of damage. For example, during local sprouting after injury, the denervated cells release trophic factors that act as a stimulus for axonal growth until reenervation of the target cell occurs, at which point the release of these substances from the target cell is inhibited. (For details regarding these and other plastic changes at the neuronal level, see Cotman and Nieto-Sampedro, 1984). Later we shall reorganize some of our discussion of learning around the concept of information processing, which should help the reader understand and organize the complex material that is contained in this section on plasticity and the following section on learning. In utilizing the information-processing concept, we are actually returning to a view employed by Freud in some of his earliest work (1891b, 1895).

McClintock (1984) in her Nobel acceptance speech described how even the genome is aware of stresses in its environment and capable of making adaptive changes (also see Suttcliffe et al., 1984). Kandel (1983) details how the changes at the synaptic level may eventually lead through identifiable steps to changes in the genome by way of effects on the operator gene, thus allowing the expression of capacities built into the genetic machinery but not previously expressed. It is becoming clearer how nature and nurture affect each other. The outlines of what learning represents in terms of changes in nervous system anatomy, chemistry, and physiology seem almost at hand.

The factors determining neuron behavior during development or following injury are complex and include both the general humoral and the cellular environment, as well as various specific trophic factors and "neurite-promoting factors"; various models of plasticity and repair are currently under investigation (Varon and Manthrope, 1985). Although it has been known for a long time that regrowth is easier in the peripheral than the central nervous system, it is now clear that this difference is not because CNS nerve cells lack regenerativity (Black, 1985). The discovery of the plastic potential of central neurons has also led to an increased interest in the role of neuroglia, which apparently is more complex than the function of "dampen[ing] the wider fluctuations in neuronal connectivity and thus ensur[ing] the permanence of the 'wiring' arrangements determined during development . . . " (Aguayo, 1985, p.37).

The study of the anatomical details of the neuroglia, for example, astrocytes that envelope and "scroll around" the cell bodies and dendrites of Purkinje, cells in the cerebellar cortex, has shown that such cells as these may very well divide neurons into functional groups and provide an organizing as well as an isolating function, thus promoting neuronal specificity. There is reason to believe that these glially created compartments are undergoing constant alteration, presumably in response to the information flow through them (Palay and Chan-Palay, 1974, p. 321). New evidence indicates that the particular type of glial cells called Schwann cells may further contribute to neural plasticity through the production and release of the glial growth factor (Brockes, 1984).

"All information processing in the brain involves neurons 'talking' to each other at synapses by means of neuroactive chemicals" (Snyder, 1984, p. 1256). Such neurotransmitters may be amines, amino acids, or peptides. Recently, over 50 peptide neurotransmitters have been identified (Snyder, 1984); they include enkephalins, vasopressin, and prostaglandins (Gasanov, 1984). These neuropeptides play a complex role in adaptive (i.e., learning processes. Oderfeld-Nowak and associates (1985) report that brain repair is accelerated by the chronic administration of a mixture of the four major brain gangliosides, and Vartanian (1985) describes how peptidelike substances appear in the cerebrospinal fluid following cortical injury, release lower brain centers from their normal place within the hierarchical organization of the brain, and produce a functional reorganization. "Thus . . . peptide factors . . . provide information on the ability of neurons at all levels to rearrange, actively, their intercellular connection. Possibly, these synaptic modifications account for [some of] the plasticity of the brain and memory processes" (Vartanian, 1985, p. 99).

At the clinical level, Bach-y-Rita (1981) writes on the relationship between plasticity and the need for complex stimulation and rich opportunities for experience, something understood a generation ago by Benedek (1937/1938) but apparently never integrated into neurology. That is, affective inputs appear to play a decisive role in the process of releasing (by selective inhibition) the adaptive changes we call learning, which occur throughout life and include recovery from neural injury (see also Gazzaniga, 1979, p. 813).

LEARNING

The enormous topic of learning cannot be reviewed briefly; therefore, the interested reader is referred to such interdisciplinary efforts as Rosenzweig and Bennett (1976), Young (1978), Simon (1979), Albus (1981), Alkon (1985), and Bignami and associates (1985). Essentially, four adult neurological mechanisms for learning are known: new neuronal branches and synapses, reverberating circuits, use-dependent modification of existing circuits, and the alteration (swelling) of pre-and postsynaptic processes (Alkon, 1985, p. 1037). Nevertheless, our general position is that *learning occurs at all levels within the hierarchically organized nervous system and that describing the plastic changes associated with interactions at each level between the biological givens and the environmental inputs constitutes the most elegant answer yet to the often posed question as to how nature and nurture interact to produce the complex changes we call learning.*

According to Snyder (1984), "the embryonic and newborn brain extends many more neural processes than will ultimately be employed by the adult organism. Subsequently, inappropriate neural connections will be pruned away by the death of nerve cells" (p. 1255). This "pruning" constitutes a major means of early brain synaptic organization and reorganization. Although some have questioned whether new neurons are produced only during the prenatal or early postnatal period (Kolata, 1984; Nottebohm, 1984), recent experiments (Rakik, 1985) suggest that this rule (of no new adult neurons) may be absolute. (See Cowan, 1979, and Cotman and Nieto-Sanpedro, 1984.) The other major mechanism for synaptic organization is based on the orchestrating effects on cell migration of neural cell-adhesion molecules (N-CAMs).

The embryology of the nervous system is not central to our discussion; however, a germane question is how the mechanisms for learning in the adult relate to those in the immature brain Nottebohm's, 1981, (See discussion of DeVoogd and Nottebohm's, 1981, work in Chapter 2.) Of particular interest is whether environmental changes and personal experience can result indirectly in changes in cell-adhesion molecules and whether these might not be mechanisms for some of the adaptive learning changes that involve CNS structures on a scale larger than individual neurons. There is certainly sufficient evidence that these molecules, or N-CAMs, contribute substantially to the way in which different cells come together to form the larger substructures of the nervous system (See Edelman and Chuong, 1982; Edelman, 1983; Hoffman and Edelman, 1983). What remains to be shown is whether these N-CAMs are also active in the continual restructuring of the mature nervous system and how this restructuring might be accomplished in terms of linking up specific environmental factors with specific changes in such variables as N-CAM sialic-acid content, structure, and so forth. (See Stallcup, Beasley, and Levine, 1983, for some preliminary support of these speculations; also see Kandel, 1983, pp. 903-904, for a review of the complexities of neuronal diversity and recognition.)

At a higher level of organization there is the question of the relation between learning and such developments as the establishment of cerebral laterality. In chapter 2, we reviewed this subject in some depth and attempted to describe some of the effects of psychoanalytic interpretations on connecting the two cerebral hemispheres, but our coverage was incomplete in a number of respects. For example, it has not been possible to describe precisely how the two hemispheres are normally coordinated in their activity because this is not yet known. Yet, clearly, this knowledge is a prerequisite for understanding how psychoanalysis may alter these coordination patterns. In the rest of this chapter we shall be seeking answers to such questions as these and attempt to explore some general mechanisms of integration within the central nervous system.

Our investigation starts with the observation that "the left hemisphere not only matures embryologically faster [than the right], but establishes connections with the subcortical areas and spinal chord in advance of the right side" (Joseph, 1982, p. 12). The implication here is that the left hemisphere assumes dominance over the motor apparatus while the right hemisphere is destined to establish its "greater abundance of reciprocal interconnections with the limbic system" (p. 16). Since the two hemispheres are poorly connected before the age of 5, and myelinization is incomplete to some extent even at 9 or 10; thus, learning during the earliest, and apparently most critical, years appears to be conducted without any obvious mechanism for the coordination of the two cerebral hemispheres. Since infants and small children do not resemble patients with transection of the corpus callosum, however, we feel that the hemispheres are coordinated and integrated in their functions from the very beginning of life, probably by a number of different mechanisms. At the very least, the cognitive style in using the two hemispheres in various tasks, either individually or in some complex patterning, seems to be learned and susceptible to change.

A consensus appears to be growing that plastic changes are potentially storable at all levels within the mature nervous system, or in at least "all systems engaged in a particular experience" (Livingston, 1976, p. 10; see also Agnati and Fuxe, 1984, p. 93). This would include the level of cytoplasmic macromolecules, intercellular organelles, synapses, individual neurons (including their critical glial environment), simple neuronic circuits, ensembles of neurons organized into modules of greater and greater complexity, larger structures of the brain, such as the cerebral cortex (hemispheres), thalamus, basal ganglia, various other brain stem nuclei, the limbic system, the cerebellum, the spinal cord, and so on. Importantly, there is a distinction between plastic change and learning. To be comprehensive we shall now try to clarify this distinction and to describe the significance for learning of changes at each level. It may eventually be possible to explain how changes at each level are linked to changes at other levels. However, knowledge of a number of the essential details regarding linkage remains missing.

To begin with, the distinction between plastic change and learning, "changes in the tendency to respond are indeed a simple form of learning, known as sensitization, which is common in simpler animals, which have no discriminative mechanism. But the memory records [i.e., the 'programming'] that we are really interested in are discriminative and *specific*" (Young, 1978, p. 83). Thus, all change can be seen as learning, but all learning does not involve complex or discriminative ability. Although there is some research to suggest that the learned behavior of flatworms can be transferred by grinding them up and feeding them to other flatworms and is therefore "chemical" (McConnel, 1962), the consensus seems to be that the more complex or discriminative type of learning has to do with circuits that first become more active by establishing some short-term memory process on the basis of a relatively brief or time-limited cycling effect and then eventually become transformed into relatively permanent or long-term circuits because they are reinforced (Young, 1978). According to Kandel (1983), "although we are beginning to understand aspects of the molecular changes underlying short-term memory, we know little about long-term memory" (p. 907) We shall be discussing learning changes of each of these sorts. Where ambiguity may occur we shall try to make clear which form of memory we are considering.

Our survey covers a variety of approaches to the subjects of learning and plasticity: pharmacological–neurohumoral, neuropsychiatric–neuroanatomical, artificial intelligence, ethological, and learning-disabilities theory. We hope the reader will obtain a view of how simple learned (habituation or sensitization) circuits become facilitated by usage and are gradually preempted into larger ensembles that subserve adaptive discriminative kinds of learning. We shall generally begin with changes at the lowest level and then proceed to higher levels of integration and coordination within the nervous system.

Kandel (1984) reviewed the history of brain-localization research, tracing the early ideas of Gall to those of Broca, Wernicke, and Merzenich, and finally to his own current work and that of others on such simple neuronic systems as the marine mollusk aplysia (Rayport, 1981; see also Scheller et al., 1984). Using the learning paradigms of sensitization and habituation Kandel and his associates map out the changes within the neuron and speculate on the connecting links to changes at the level of the genome by means of the operator gene. The steps initiated by nervous activity in the single neuron as part of a learning circuit of any complexity involve the following sequence or "cascade": entrance of a neuroactive substance into the neuron (e.g., the neurotransmitter 5-hydroxytryptamine, also known as serotonin), alteration in intracellular ATP, increase in cyclic AMP as ATP breaks down into cyclic AMP and pyrophosphate, the activation of ionic channels for calcium, the alteration of the operator gene, and the opening up of parts of the genome previously unexposed for possible expression.

Summarizing complex mechanisms runs the risk of providing too little for some and too much for others. We are trying to run a middle ground, describing without oversimplifying. Those interested in the details of the intracellular chemical control mechanisms mentioned in the previous paragraph should consult Kreiger, Brownstein, and Martin (1983) or Alkon (1985); those wishing to pursue further details of the calcium-related calmodulin memory mechanism alluded to earlier should consult Lynch and Baudry (1984).

Regarding a higher level of the brain's hierarchical structure, Kandel (1983) reports that the right hemisphere possesses areas that are the analogues to Broca's and Wernicke's areas (of the left hemisphere) and that appear to subserve similar functions but apparently involve the emotional (prosodic/rhythmical) or affective components of motor and semantic speech, respectively. This understanding about analogues shows that the right hemisphere and the limbic system form a major subsystem of the brain subserving emotion.

Some clinical material will help demonstrate how much of what we already know from the classical psychoanalytic approach dovetails neatly with, and can be explained simultaneously by, the kinds of knowledge from neuroscience and the ancillary fields just described. Psychoanalysis clearly stands up to this kind of testing. Later we shall return to a more detailed consideration of what learning might be, from the interdisciplinary viewpoints mentioned earlier, and attempt further to integrate the perspectives of psychoanalysis and the other brain sciences.

CLINICAL VIGNETTES

Mr. B., a 37-year-old intellectual, entered psychoanalysis because of mood swings that initially raised the possibility of manic-depressive illness in the mind of a consulting psychoanalyst. A more detailed workup, including lithium-efflux testing (as a research protocol), as well as continued observations, however, challenged this diagnosis. Prominent depressive mood swings made it difficult for the analysand to function on his job in an investment banking enterprise, and he frequently found himself paralyzed by the inability to decide whether to remain with this firm or seek employment elsewhere. Intellectually, he appreciated that his job represented a rare opportunity, but some unknown factors prevented his making any sort of useful analysis of the situation. His major and frequent complaint, which we shall attempt to understand from a multidisciplinary viewpoint, was: "I feel like I just lost 20 IQ points!"

Early in the analytic work, Mr. B. and the analyst established a rapport that provided some significant insights into his dilemma; nevertheless, his mood continued to be too unstable for him to do reliable work. For example, at one point we recognized that daily he was reliving in his mind the events and affects of his father's hospitalization for a terminal illness several years earlier. The discussion of the details of this loss began to alter his helpless state and led to the recognition of other anniversary reactions that he could see were part of an attenuated mourning process. This ushered in a phase of subtle but intense transference reactions and resistances that occupied the work of several years and led to a consideration of some character pathology that preceded the parental loss. Instead of attempting to delineate the complex features of this analysand, we shall select some of these transferences and the technical strategies that seemed to prove useful. These will be presented along with some partial explanations, first in terms of the usual analytical viewpoints and then in terms of possible brain mechanisms. This, of course, is only meant as a speculative enterprise; nevertheless, we feel the mechanisms invoked are in principle testable, if not now then in the very near future, using noninvasive techniques (see chapter 2; Phelps et al., 1985; Greitz, Ingvar, and Widen, 1985).

Mr. B. would frequently interrupt his free associating with what could have been a pernicious resistance if not impeded: "But what's the point?!" Then he would lapse into a gloomy silence, feeling more and more helpless. In each case, the analyst reacted differently, but gradually one approach appeared to be more effective than others. Mr. B. was simply reminded in a nonintrusive and gentle way of what he had been saying just before he said this. At times, not only would he be reminded of the recent comments, but a speculation might also be added that suggested possible connections to previous material either in the same session, the previous session, or both. (The reader will understand this as an attempt to interfere with a number of kinds of resistance but in particular with the use of repression and disavowal, as well as a number of other defensive maneuvers, such as isolation of affect, intellectualization, displacement, etc.). The patient would then remember what he had just forgotten about in the earlier material and begin to acknowledge its possible significance emotionally.

A second strategy that proved highly effective in dealing with the silences was for the analyst to allow himself, after a while and when nothing else seemed to work, to verbalize such things as particular songs that might then be running through the analyst's head in response to what the analysand had just been saying. These pump-priming associations seemed frequently to set off a rich tapestry of associations and still more complex transference relations, and they greatly increased the analysand's interest in and progress within the analysis. The question is, why did these various strategies work?

In chapter 2 we state that the two hemispheres appear to be functionally disconnected during the use of repression and disavowal and that transference interpretations appear to bridge the hemispheres. This hypothesis is supported by such disclosures as this patient's feeling that he was at times losing IQ points (possibly because of some interhemispheric disconnection, although one could assert that the problem was chemical); however, we do not feel that this explanation goes far enough. What the "pump priming" seems to accomplish, along with the transference interpretations, is to reduce the resistance to a particular train of associations (It also familiarizes Mr. B. with the analyst's style of using his two hemispheres.) But why is this so? One interesting possibility is that when the patient hears the associations of his analyst, some specific memory system is primed or evoked. We shall discuss this more fully presently (see our comments on Winson's work), but for now we wish merely to mention the general possibility that interpretations might effect "neuronal gating" within critical brain stem nuclei or within the hippocampus in particular. If this is true, then this "gating" control may be what allows the transference interpretation itself to be fed into a system that would be functionally enlarged because the hippocampus (limbic system) and the cerebellar system memory mechanisms would now (without the "gating") be linked as inputs to the total system; that is, the gated brain stem nuclei would no longer block out specific limbic and/or cerebellar or other inputs. Before the gating was interfered with (e.g., during anxiety), these systems would very possibly have been knocked out. A second possibility is that the patient's style of coordinating his hemispheres (and, of course, the analyst's style of using his brain) has been altered by his awareness of the analyst's style (see Schlessinger and Robbins, 1983, p. 9). This second possibility raises the question of how the hemispheres are coordinated in their activity, a topic that we expand on later in this chapter.

Mr. H. is a 46-year-old French-horn player with some learning disabilities. He came into treatment because he and his referring psychoanalyst had been unable to terminate after more than 10 years. Because of the failure to negotiate this impasse, it appeared to the patient that nothing positive had been accomplished although many gains had actually been made. The analysand was regressed, and anger colored most of the early work; only gradually did he trust his new analyst, and even after two years he seemed to continue to expect the imminent repetition of the previous "rejection".

In keeping with our theme we would like to focus here on some details of the interchange that may throw further light on the hemispheric-blocking mechanisms that we are struggling to understand. In this regard one observation seems to stand out. The analysand needed his analyst to realize that if he assumed a strange posture or position during a session, this should not be taken ipso facto as evidence of insanity. Needless to say, the first matter investigated was why the patient had this particular fear. This led to the discovery and exploration of what appeared to be memories of some truly bizarre behavior on the part of the parents. The father apparently had engaged in incest with the analysand near puberty or possibly earlier and seemed to have undergone a psychosis of a paranoid kind, most probably during the patient's early childhood. In addition, the mother gave birth to a baby brother when the patient was two and was remembered to have no longer involved herself emotionally in the life of the patient; in particular, she apparently stopped what little touching of the patient she had formerly allowed herself.

After he assured himself that it was safe, the analysand on several occasions allowed himself to squeeze into the corner of the room, between the analytic couch and both walls. From this vantage point he could see the analyst and yet not risk being attacked from the rear (or so it seemed to the analyst). He could then continue to associate to some particularly painful affect states that seemed to involve more detailed memories of either being touched inappropriately by the father (and a brother) or not being touched by the mother. These memories appeared to be available partly because of a general reduction in anxiety, but also seemed to be linked somehow to his assuming this particular posture.

Another clinical observation: Mr. H. felt particularly soothed by any attempt by the analyst to communicate musically, that is, by humming a tune or singing words, to remind the patient of some important feelings or associations. More about this later.

From the neurological perspective, it is interesting to ask why the posture described earlier was required. It would surely be easy to stop at the obvious explanation that the analysand was protecting himself from an anal (homosexual) penetration, which was certainly true. However, if the impression is correct that the posture had additional meanings as well, then we might wonder again if some particular memory system would be invoked by such a maneuver, and if so what this would be. To us this suggests the vestibulocerebellar system, which is specialized to deal with information coded into body movements, and we shall pursue this subject in more detail later. For our purposes here, however, we wish also to note the possibility that in this patient the bridging of hemispheres accomplished by some transference interpretations may also invite another kind of bridging to complement it, as it were, namely, a bridging of the hemispheric systems (with their intimate and asymmetric linkages with the motor system and with the limbic system) with the vestibulocerebellar system. It is the latter system, for example, that probably involves the earliest memories of the so-called body/mind/self, and it stands to reason that the recall of such early "rhythms" might be enhanced by the vestibulocerebellar system working in concert with the other major memory systems of the brain. The assumption that bridging of the hemispheres was critical in the recovery of this patient seems supported to some degree by his love of music (particularly opera), which we feel had come as a lifesaver for him, partly because it allowed him to make use of his genuine musical ability (thus enhancing his self-esteem) and in part because this allowed him to use his hemispheres together as he combined words and music in meaningful ways in the various musical pieces that he loved. More specifically, his interest in opera probably helped him move toward partially overcoming the repression and disavowal (which we have suggested previously may very well turn out to be defensively determined interhemispheric blocks) that he had instituted at the time of the incest but that left him with a sense that he had lost the lower part of his body (including his genitals) as a part of his body image. It was only gradually yet steadily that "bridging" interpretations of the transference (which brought the hemispheres together by simultaneously undoing both specific repressions and disavowals) allowed him, for example, to regain a sense of "ownership" of the genital part of his body.

Mrs. F. is a 36-year-old mathematician who entered analysis because she was feeling like a "prop" for her husband and was suffering a kind of existential despair over ever altering this situation. We would like to focus on one aspect of this patient's intrapsychic life: what we came to call her "big limb feeling." This feeling appeared to be lifelong and seemed most closely associated with nonspecific stress, at least at first. Later on, through a series of deductive inferences and specific memories, we were able to identify this feeling as very likely a memory of her mother's psychosis (paranoid type) following the birth of the analysand's baby brother when she was two and a half. Mrs. F. recalled that her mother spoke of her mind "rotting," and the father unfortunately was not available at this time for mother's support. The analysand felt that she had to support and be a "prop" for her sick mother.

Frequently this patient seemed to be blocking affects from their associated personal meanings. Many transference interpretations ferreted out specific fears and recollections confirmatory of our reconstruction.

The focus was more and more on the feeling of limb distortion, which seemed to be the patient's recollection in body language of the mother's earlier psychotic affect states. The analysand did not in any way suffer from a psychotic or schizophreniform illness herself, but she was able to know in part what this might be like because she had in her brain the recollection of her perception (circa age two and a half) of how her psychotic mother had felt.

Almansi (1983) and Blum (1977) have described the recovery during analysis of very early memories and affect states, including the experience of such things as the mother's breasts, face, and hands during the first year or two (Almansi, 1983, pp. 392, 410–414). We shall not linger on this subject of early recollection except to note that it suggests to us that during analysis some bridging may occur to both cortical and also possibly much earlier memory systems. We speculate that cerebellar recall might be involved. The cerebellar would assist the memory of the cortical and other brain memory systems (hippocampal or diencephalic) in keeping track of the earliest rhythms of life. Remember that before the corpus callosum of anterior commissure begins to myelinate (after age three and a half) the cerebellum is present as a major memory link with coordinating, integrating ties to both hemispheres.

PHARMACOLOGICAL–NEUROHUMORAL APPROACHES TO LEARNING

Having already seen the body of clinical material with which we are concerned, the reader will better appreciate the challenge of attempting to understand it from multiple simultaneous perspectives, including some from other brain-related disciplines. We shall now proceed with our review of some work in the area of learning theory, starting with pharmacological and neurohumoral perspectives. Where possible we shall make additional comments on the clinical case material previously presented, although being comprehensive is only a theoretical possibility. We shall be satisfied, therefore, if we merely gain some additional insights into possible learning mechanisms.

Immunology has contributed greatly (for example, through its methodology) to the identification of the brain's major neurotransmitter systems (Reichert, 1984). Complicating the situation, however, is the fact that both immature and adult neurons are fully capable of changing the kind of neurotransmitter they release, thus adding "a new dimension to our appreciation of neural plasticity" (Black et al., 1984) and an additional complexity to the problem of diagraming the functional circuitry of the brain!

We begin our discussion with the special class of neurotransmitters known as monoamines (norepinephrine, serotonin, and dopamine), which appear to have gotten particular research attention. Winson (1985) points out that, unlike other neurotransmitters that produce action potentials in target cells directly, "the monoamines appear to modulate the effect produced by other neurons that also impinge on these same target cells" (p. 193). Winson has studied the monoamine systems within the brain stem, his particular interest being in how neuronal gating within the hippocampus affects the limbic system's role as a central processor of memory and emotion. He concludes that hippocampal electrical "gates" are controlled by brain stem nuclei; in particular, he singles out such input as brain stem norepinephrine from the locus ceruleus, studied extensively by Floyd Bloom of the Salk Institute.

Although there is some debate about its exact role (Routtenberg, 1976), the locus ceruleus becomes important for a discussion of discriminative learning because there is evidence that it may be necessary for long-term storage of engrams in parts of the brain (brain stem, hippocampus, cerebellum). The conversion of circuits from short- into long-term memory-storage form appears to occur when the locus ceruleus provides input to the relevant circuit. That is, the locus ceruleus "is connected with the reward, or positive reinforcement, system of the brain" (Gilbert, 1975, p. 698) and thus apparently "knows" when to tip the scale in favor of reinforcing those circuits involved in behavior that seems adaptive in terms of pleasure/pain experience (Gilbert, 1975, 1976). More about this later when we discuss some specific mechanisms of cerebellar learning. (Obviously we feel that the cerebellum is important in learning, but this topic remains controversial; for example, Shepherd, 1979, pp. 245–246 might be consulted for criticisms of the idea that the cerebellum is involved in learning, at least to the same extent that the hippocampus and neocortex are through their particularly "labile synaptic actions.")

Winson (1985) wonders about the exact role of the hippocampus: whether the hippocampal circuitry is "functionally modified during certain activity states . . . [by brain stem input, where norepinephrine is the agent] so that specific information processing function[s] [are] performed during these states" (p. 201). He is specifically interested in understanding the significance of the REM/nonREM phenomena, which are for him markers of critical information processing of the right and left hemispheres, respectively, occurring "off-line" (pp. 206–207). In his opinion this off-line processing saves the prefrontal cortex from having to process as well as integrate new information simultaneously (p. 207).

In this regard it is interesting to speculate whether such experience as Mr. B.'s felt sense of losing IQ points might alternatively represent his attempt to process new information on-line, versus off-line. It is not inconceivable that psychoanalysis, by its exhortation to the analysand to take his time thinking about conflicts, invites him to shift from an inherently difficult (on-line) to an apparently easier (off-line) processing mode.

Winson's speculations make sense to us, since we are concentrating on the problem of hemispheric coordination; we are, however, particularly curious about whether or not the cerebellum might be one of the lower centers that is providing critical "gating" functions within the hippocampus via its interaction with various brain stem nuclei. Alternately, we wonder whether the brain-stem nuclei that gate the hippocampus may not also gate the cerebellum, thus interfering with the cerebellar input to the cerebral cortex. The significance of this will become clearer later when we develop our ideas regarding how brain-stem nuclei may play a more general role in bringing together or disengaging major subsystems of the brain (see also Kitney et al., 1984, p. 368, in this regard). Also, as noted in each of the cases presented, there is the possibility that some of the mentioned psychoanalytic interventions that were required, for example, in order to allow certain transference interpretations to work, accomplished their results by means of brain-stem and/or cerebellar alterations. The neurophysiological mechanism we and Winson are considering in the brain stem and the cerebellum would constitute the basis for such "preparatory" work within psychoanalysis that theoreticians such as Gedo (1979) have identified as "beyond" (meaning preparatory to) interpretation.

A second category of chemical approaches to learning involves the work on endorphins, whose role is as yet uncertain but that appear to be modulators of the hormonal category of neurotransmitter and at least a partial explanation for so-called placebo responses (Cleghorn, 1980, p. 183). According to Rose and Orlowski (1983), these substances are concentrated in areas of the brain that involve hormone release, pain perception, cardiac function, respiration, appetite, mood, and learning (p. 131). That the morphine antagonist naloxone given alone in physiological concentrations alters the efficiency of learning suggests that endorphins play some role in learning and memory processes (p. 133; also see Levin, 1985). Whether this involves effects on the value of reward and/or punishment, attention, memory consolidation, or retrieval is not clear, nor is it clear what anatomical systems are particularly involved in these effects.

A third chemical approach to learning involves investigation of cholinergic mechanisms. Overstreet (1984) discusses attempts to correct memory deficits in aged humans by manipulating the cholinergic system. Here he relies on Drachman's observation that the effects of scopolamine on memory seem to parallel what is seen in senile memory changes (Drachman, 1977). Apparently, the cholinergic system has no specific effect on habituation (Overstreet, 1984, p. 135), although the fact that aged humans with memory deficits benefit from piracetam, which increases cerebral oxidation if it is given with the acetylcholine precursor choline (but not apparently if it is not), suggests that the cholinergic system is capable of playing a role in learning (p. 145; also see Bartus et al., 1981). The consensus seems to be, however, that other neurotransmitter systems in addition to the cholinergic one are involved in learning and memory, particularly the noradrenergic (monoamine) neurotransmitters (Overstreet, 1984, p. 138).

NEUROPSYCHIATRIC–NEUROANATOMICAL APPROACHES

Attempts to localize changes in learning, which include those of Lashley (1950) to discover an engram in maze-trained rats, appear to come to fruition in the work Merzenich and associates on monkeys given learning tasks (Kandel, 1983) and experimental amputations Merzenich et al., 1984). It is possible to show that with increased use of a digit the somatosensory map of the monkey cortex changes, with the area representing the finger that is involved in learning growing in size! In the amputation experiments, it was shown that when a finger is amputated the cortical cells learn about the change and the cortex somatosensory map adjusts so that the area of cortex previously representing the amputated finger now represents the fingers on either side of the lost digit! On the basis of these findings there can be no debate that some changes with learning occur at the cortical level and relate to the plastic manner in which cortical sensory cells represent a given sensory field.

A discussion of learning and memory would be incomplete without a comment on the work of Penfield and Mathieson (1974) on the limbic system, specifically on the role of the hippocampi. Expanding on Penfield's earlier work (1958), they report on a disastrous loss of memory in two patients who required left-temporal lobectomy for intractable epilepsy. What was not appreciated presurgically was that each patient already was functioning without a right hippocampus because of injury at birth. When the remaining (left) hippocampus was removed, the patients ability to scan and call to mind past experiences was drastically reduced (Penfield and Mathieson, 1974). Penfield concludes from this study that each hippocampus has "access keys" to a critical recorder of events in the brain stem (diencephalon), which appears to store memories in longer-term storage than the long-term stores of each hippocampus (p. 153). He also points out the asymmetries between the two hippocampi: the left hippocampus specializes in the recall of verbal experience; the right specializes in scanning episodic memory, and orientation in space (p. 152). There is thus clear evidence that the cortex of the two temporal lobes works together with the two hippocampi and the brain stem in serving particular aspects of memory.

Rozin (1976), in reviewing neuropsychiatric research, starting from Jackson's (1958) ideas on the brain hierarchical organization and moving to contemporary work on artificial-intelligence models, has chosen to focus on amnesic syndromes as a way of finding a key to puzzling memory/learning phenomena. These syndromes, such as Korsakoff's, described in 1889, include (1) the failure to recall and recognize recent events (meaning within the last 15–20 seconds); (2) loss of the sense of familiarity and personal reference (i.e., episodic memory); (3) at least some storage of recent events; and (4) particular difficulty with remembering completely new material, such as personal names (Rozin, 1976, p. 7). Rozin reviews the data bearing on theories that divide human memory into three stages: (1) a sensory stage; (2) a limited-capacity, short-term memory stage that seems to depend on some recirculation mechanism for retaining material; and (3) long-term memory, which represents a totally different system, one with high storage capacity but with much slower access rates (p.9). Rozin contrasts Geshwind's work with that of Luria; the former is characterized as seeing the brain as divided into neat elements that fail to function when specific processing centers are disturbed or disconnected, whereas the latter is described as seeing the brain as not so neatly divided but instead as organized around zones of progressive elaboration (p. 9; cf. Marr's work on brain modules, referred to in the next section). Geshwind and Luria seem to agree that the inferior parietal lobe of the cortex is critical for the integration of the different sensory modalities. Sensory integration also occurs, however, at other levels, such as within the limbic system and in the cerebellum.

According to Rozin (1976), the Korsakoff syndrome occurs because of bilateral damage to the mammillary bodies, which are apparently especially sensitive to thiamine deficiency (p. 23). These are parts of the limbic system, which we have noted is a major subsystem of the brain subserving both affect and memory. Bilateral damage to other areas within the limbic system—the anterior nucleus of the thalamus, the fornix, or either hippocampus—apparently also results in an amnesic syndrome (p. 24), in particular one involving episodic memory (see our discussion of interhemispheric blocks in chapter 2). This suggests strongly that memory deficit is a retrieval block and reminds us that in general perhaps the most critical phase of memory is retrieval, which appears to be particularly susceptible to disturbance. Later in this chapter we will draw on this observation to elaborate on what we feel may be an important contribution of the cerebellum to limbic system and overall CNS memory retrieval. Also remaining to be answered are such questions as how retrieval is facilitated or obstructed by particular learning strategies.

Gazzaniga (1976) reminds us that the brain stores memories at multiple sites and in multiple ways (p. 57). In this respect the right cerebral hemisphere is "an equal partner [with the left] in the cognitive abilities of everyday experience" (p. 62). Most interestingly, he reminds us that information transfer within the brain occurs when one part of the brain observes the output of other parts in terms of overall behavior. In the psychoanalytic situation, of course, this phenomenon is exploited enormously. Right–left functional disconnection syndromes show the criticality of the interaction of such brain subsystems (Arbib, Kilmer, and Spinelli, 1976, p. 124). Moreover, such structures as the hippocampus (and the cerebellum, discussed later) show how "an abundance of simple, randomly connected components with plastic synapses [can] be put to powerful use in brains as simple classifiers of past experience" (p. 126).

The model of the storage of key programs at multiple locations (i.e., within multiple memory systems) makes a great deal of sense, and this is probably why, for important matters, we try to involve as many different sensory modalities and learning strategies as possible at the time that some particular learning occurs. Consider, for example, the strategy that proved effective in the cases of Mr. B. and Mr. H., which included references to specific melodies and songs. The chief advantages of multiple storage are (1) protection against the loss of some vital engram, and (2) ease of recall (because the programmatic memory would be filed under and presumably retrievable by means of a number of different classificatory schemata). More about this later, when we discuss the cerebellum specifically.

Another body of neuropsychiatric data bears on the problem of understanding our "plastic" learning capabilities, and this is the data from ablation and stimulation experiments. The Tulane group, under Heath (1977), pioneered the study of the relationship between emotion, activity in the deep cerebellar nuclei, perception, and learning. Riklan and his associates (1978) and Cooper (1978) extensively review the role of the cerebellum. Ricklan and his associates (1978) believe that although "studying the cerebellum in various mammals, by stimulation or ablation methods . . . does not lead to a cohesive concept of [the cerebellum's] role," nevertheless there is reason to believe that the cerebellar input is important in the following behaviors: perceptual, perceptual-motor, emotional, activation, and arousal (p. 164). Moreover, there is no question that *stimulation of the cerebellum does appear to alter learning* (p. 168). The stimulation was provided as part of the attempt to treat intractable spasticity and/or seizures. Long-term patients showed clear improvement in their intellectual functions and in memory, as well as in accuracy on perceptual tasks (pp. 176–177). Ricklan and his associates speculate that this learning enhancement might be related to a cerebellar role in visual perception (p. 177), although it should be clear that interpreting such data is difficult to do with confidence. We do agree, however, that the cerebellum needs to be explored to better understand its role in learning.

OTHER APPROACHES TO LEARNING

In addition to chemical, anatomical, and physiological viewpoints, our understanding of learning has benefited from three additional perspectives: ethology, artificial intelligence, and learning-disabilities research. We shall now sample some relevant aspects of each of these areas. (Other areas, for example, some of academic psychology, are relatively shortchanged in our discussion, which focuses primarily on neurological perspectives. We hope eventually to be able to correct this situation.)

In recent years learning theory has been greatly expanded by work in the field of artificial intelligence (A.I.). In his monumental work on vision Marr (1984; see also Rosenfeld, 1984) begins with two questions: the what and the how. Regarding what the nervous system of a given species is designed to accomplish, the evidence is strong that all animal nervous systems are designed to facilitate survival of the organism through adaptation (cf. Hartmann, 1939). To accomplish this end, the "plastic" property of the nervous system is exploited to capture information about the environment in the form of learning. Such patterns as habituation and sensitization evolved as sophisticated but simpler modes of learning; with the evolution of the mammalian brain, complex discriminative kinds of learning appeared.

According to Marr, the how question is best answered by breaking down the brain's actions into modules, which accomplish parts of tasks in assembly-line fashion. For example, before an image even leaves the eyes much data analysis occurs, so that the computational task of the cortex is greatly simplified. (For details of this fascinating story, see Marr, 1984; Rosenfeld, 1984; and Kent, 1981.)

Albus (1981), Pearl (1984), and Simon (1979) have presented various complementary approaches to the problem of designing mathematical and/or machine models replicating human-brain-like behavior. Albus (1981) quotes the late Alan M. Turing as follows regarding the history of A.I.:

> We may hope that machines will eventually compete with men in all purely intellectual fields. But which are the best ones to start with? . . . Many people think that a very abstract activity, like the playing of chess, would be best. It can also be maintained that it is best to provide the machine with the best sense organs that money can buy, and then teach it to understand. . . . This process could follow the

> normal teaching of a child [cf. how the computer HAL is taught in the movie *2001 Space Odyssey*]. Things would be pointed out and named, etc. Again, I do not know what the right answer is, but I think both approaches should be tried [p. 5].

The field of artificial intelligence began with Turing's first approach, although the second approach—what Albus (1981) calls "the attempt to reproduce the control functions and behavior patterns that exist in insects, birds, mammals, and primates" (p. 5)—seems to be gaining ascendancy now. We agree with Albus that this learning-oriented approach will become critical for the testing of our understanding of fundamental mechanisms of CNS control and regulation.

In a nutshell, some workers in A.I. believe that some computers can be called intelligent because they learn how to learn (Alexander, 1985, p. 144). This genre of A.I. work includes such programs as Eurisko by Lenat of Stanford University. Eurisko is reportedly capable of "thinking about its own thinking, employing processes not unlike introspection and the stream of consciousness that occupies the human mind nearly all of the time" (Alexander, 1985, p. 144). Clearly, mathematicians and engineers are attacking the problem of learning about learning and are including the provision for thought experiments of an apparently human sort so that a programmable machine can learn from experience and a kind of self-reflection. We defer a fuller discussion of the relevance of work in A.I. for psychoanalysis to a later time, because space precludes our doing justice to this subject of critical importance. Readers interested in medical artificial-intelligence work may wish to consult Clancey and Shortliffe (1984). Those seeking more comprehensive general information will prefer Cohen and Feigenbaum (1982) or Charniak and McDermott (1985). The best mathematical approach of the subject seems to be Anderson and Rosenfeld (1989).

Regarding ethology, Lorenz feels "there are inherited fixed action patterns of motor coordination and an inherited responsiveness to particular stimulus patterns or releasers; learning operates so as to modify the times and contexts at which these inherited elements come into play and increases or decreases the range of stimuli that are effective" (cited in Manning, 1976, p. 147). Unfortunately, we do not know enough about the "releasers" for human learning capacity, in particular how they unfold over time. The probability is that just as with children who experience certain specific "learning windows," say, for learning how to read, which, it is hoped, are responded to by the environment in a timely way, so also later in life we continue to enter periods that are optimal for certain kinds of learning and that are expressions of an unfolding genetic plan. A number of researchers in neurology are working in this area (see Aschoff, quoted later in this chapter), and we await their discoveries, particularly those relating to entrainment phenomena and biological rhythms.

Manning (1976) claims that although habituation can be related to neural events of the type described without much trouble, "no such secure relationship is yet possible for associative learning" (p. 149). This assertion might be contested now in the light of the work of Merzenich and his colleagues (1984). The ethological approach does, however, highlight, that *learning is always species specific*. This specificity we presume, is because as one advances up the evolutionary ladder what changes is the very plasticity of the nervous system itself. Thus, for example, the simpler nervous system of insects is not as able to modify itself; under stress it responds only as it has been programmed. In humans, in contrast, stress results in more changes at more levels within the nervous system so that even stresses that seem not anticipated in the original design can be adapted to if there is sufficient time. In fact, it appears to be this very plastic aspect of the human nervous system that adds extraordinarily to its complexity.

Summarizing a conference on learning disabilities, Spreen (1976) concludes that, although "solid data is hard to come by" in this area (p. 446), there is evidence that children with learning disabilities constitute at least 10% to 15% of the general population (p. 448). The likelihood is that, untreated, a significant number of these children will become learning-disabled adults. How to classify their deficits seems much more controversial than that the deficits exist and cause substantial dysphoria in individuals and families. Etiological factors are even less understood. Disruptions are imagined at many levels within the brain: for example, in the reticular activating system or midbrain level (to explain those cases where the attention-deficit disorder seems primary; see p. 454); at the neocortical level, either on the basis of unilateral malfunction, such as Geschwind's suggestion of late myelinization of the parietal lobe (p. 458), or some difficulty in hemispheric collaboration or dominance (pp. 462–464); and at the level of specific perceptual systems controlled by the brain stem and vestibulocerebellar mechanisms. In this regard, Leisman's findings (see Spreen, 1976) may be significant in that some learning-disabled children (with attentional and reading problems but without demonstrable brain damage) resemble hemiplegics and brain-damaged children in showing disturbances in saccadic eye movements, which are required for normal transmission and processing of information (p. 464). Miles (1980) has described how this oculomotor system is actually composed of five independent systems, each of which could presumably fail, either alone or in combination: (1) a saccadic system that transfers gaze to new objects while keeping the foveal image clear for inspection; (2) a foveal pursuit system concerned with tracking in the event the object should move; (3) the vergence system, which ensures that both eyes stay aligned on the object of interest; (4) the vestibulo-ocular reflex, which maintains stability of the retinal image by adjusting compensatory rotational movements of the eyes to compensate for rotation of the head; and (5) an optokinetic system, which tracks and reduces residual retinal slippage when the other systems fail (p. 317).

It is worth going into detail here to help the reader appreciate the advantages to psychoanalysis of taking into account the explosion of knowledge in neuroscience. For example, the general feeling that we are often dealing in an analysis with learning disabilities seems to be an insight that needs to be pursued, and this does not mean that we are talking about anything other than functional brain disease. Hartmann (1939) certainly made clear how ego functions can develop either with "primary" autonomy or "secondarily" after the mastery of a conflict that disturbs their expression. In fact, this view is what we had in mind when we discussed in chapter 2 how the analyst, in assisting with the resolution of character defenses that involve the use of repression and disavowal, is very possibly simultaneously addressing a kind of learning disability that represents a functional interhemispheric communication block. This kind of block is undone by the kinds of transference interpretation that tend to translate the output of one cerebral hemisphere so that it can become the input of the other. Clearly, other kinds of learning blocks might be expected to require different kinds of interpretive intervention in the psychoanalytic situation.

The topic of learning blocks in psychoanalytic patients will generate considerable controversy because it is likely that learning blocks result from a number of different etiologies, and the possibility of making an error in diagnosis is considerable. For example, Sklansky (1984) and Hoit (1984) write cogently about learning blocks that seem secondary to conflicts with parents or that have resulted in "a conflict-laden intrapsychic state" in the analysand (Sklansky, 1984, p. 213). That is, these authors tend to see learning blocks as representing significant self-psychopathology. However, we appeal to the reader to keep an open mind since in some instances, the frequency of which is difficult to determine at present, the *self-pathology may not be the cause but, rather, the result of a learning block* (e.g., stemming from a processing problem within any of the many brain systems for eye tracking). There appear to be situations in which the boundary between "organic" and "psychological" disappears either because a phenomenon fits both perspectives simultaneously or because it fits no model without procrustean efforts. Psychoanalysts, we feel, need to be particularly careful in labeling learning blocks as psychological; the history of our work with schizophrenia and with autism should have taught us something. First parents were blamed; only later were biological factors identified in some cases; and finally psychoanalysts made an attempt to achieve a more balanced view, taking into account and weighing all the variables in each individual case.

CNS INTEGRATION AND COORDINATION: A ROLE FOR THE CEREBELLUM

The second part of this chapter, although somewhat speculative, will connect with the first part and amplify our feeling that there is a critical, unmet need for a psychoanalytic learning theory that can be integrated with what is known about learning phenomena as studied by other disciplines. In our attempt to articulate some of the regulatory and organizing principles that learning seems to involve we now need to review some additional neurological knowledge.

Among the major subsystems of the brain there are a number of candidates for the role of coordinator and integrator of its overall activities. The prefrontal portion of the cerebral cortex has long been recognized as the part of the brain that involves the planning and initiation of activities at the highest level. In addition, since Papez's pioneering work on the limbic system, this system and its hippocampal and other components have been understood to play a major role in the overall storage of memory, integration of intersensory experience, and coordination. Eccles and his associations (Eccles, 1979, 1982; Eccles, Itoh, and Szentagothai, 1967; Itoh, 1984) have suggested that now the cerebellum needs to be added to the list of candidates for playing an integrating or coordinating role. In this section we explore some of the evidence for a cerebellar contribution to higher brain function.

A number of observations make one begin to consider a role for the cerebellum in affective learning processes (see Thompson, 1986). Each observation by itself would be "innocent" enough, but taken together they become more weighty evidence. First, the cerebellum is a subsystem of the brain that receives input from and sends output to every other part of the nervous system (Eccles, 1979; Frick, 1982). (We are aware that the cerebrocerebellar connections are made through brain stem nuclei, but we do not feel that this technical point invalidates the contention that the cortex and cerebellum are intimately related). On this basis it seems possible that the cerebellum is involved in part of the overall orchestration of a number of nervous system activities that range widely. These include the major affective elements within the brain (prominently the limbic system and thalamus), the motor system (of which it is the principal regulator), and the brain's systems of integrating sensory modalities of every kind (which keep the cerebellum in continuous touch with both the external and internal milieu).

Second, the cerebellum contains more neurons than does any other part of the nervous system (Eccles, 1979; Shepherd, 1979), which gives it the "computing power" to contribute to the integration and coordination functions that we assume human learning involves. Third, any system that would play a critical role in our emotional lives should contain prominently a representation, or its equivalent, of the body/mind/self, and this most certainly is a property of the cerebellum, which contains several representations of the body (Thach, Perry, and Schieber, 1982). The cerebellum is also particularly constructed so as to keep track not only of the body parts and their relationship to each other, but also of posture and acceleration.

Fourth, not uncommonly over the course of an analysis significant, permanent improvements in the gracefulness and coordination of the analysand can occur, and this implies to us a possible cerebellar contribution to the learning that occurred during the treatment process, as well as the general result of a diminution of anxiety. Fifth, it is our impression (and we believe it was Freud's when he first described empathy) that the analyst, in generally attempting to understand, will automatically or spontaneously assume at critical times the same posture as the patient (and vice versa), in great detail, and we feel this implies that the cerebellum is involved to some degree in communicating and gaining insight into or access to our feelings and memories, which may well be organized around action patterns, postures, or particular sensory modalities (see Levin, 1980) Last, we feel that part of what we call good mothering (soothing activities) involves certain repetitive and rhythmical stimulations, which suggest to us again an important contribution of the cerebellum.

In analyzing comic movement, Freud (1905) states that he has observed that when people express such attributes "as largeness or smallness, they "follow a kind of 'ideational mimetics." The person is "not content to make his idea plain . . . by the choice of clear words, but that he also represents its subject-matter in his expressive movements: he combines the mimetic and the verbal forms of representation." That is, the person "demonstrates quantities and intensities" so that a high mountain or short person is denoted by a raised or lowered hand. Freud adds that if the hand movements are inhibited through learning, then the speaker will raise or lower his voice. Moreover, if this voice modulation is self-controlled, then the speaker will substitute something like opening his eyelids largely or closing them tightly when it comes to the large or small imagery, respectively (pp. 192–193). Clearly, both hemispheres are making their unique contribution (a kind of dual processing).

In the same section Freud describes empathy as the understanding or apperceiving of the feeling state of another by means of putting one's body through the same movements and forming an idea of their "energy expenditure" (pp. 193–194): I behave exactly as though I were putting myself in the place of the "person I am observing . . . . In doing so I disregard the person whom I am observing and behave as though I myself wanted to reach the aim of the movement" (p. 194). To Freud, the key to empathic understanding is thus a comparison between two movements, the other's and our own. Later in this chapter we describe what this might involve in the views of the two cerebral hemispheres, possibly as coordinated through the cerebellum.

A comprehensive review of the complex subject of the ego and the cerebellum has already been done by Frick (1982). Psychiatric symptoms and cerebellar pathology have also been reported by Hamilton, Frick, Takahashi, and Hopping (1983). Our review of the cerebellum, which follows, concentrates on some of the research that bears on the plasticity and learning of this organ. We describe how the cerebellum may make a unique contribution to organizing, integrating, and coordinating the nervous system. This area has been generally overlooked, underemphasized, and rarely discussed.

FURTHER THOUGHTS ON CEREBELLAR INFLUENCE AND BRIDGING OF THE HEMISPHERES

It is well known that analysands undergo regression. Although this is a product of a number of complex variables, one variable may be the effect of generally reducing the input of the cerebellum, which, as we and Frick (1982) have indicated, includes somatosensory input, input from the reticular formation (arousal), motor input, and input from every other part of the nervous system. It is possible that what we call regression is a "decomposition" that results from the loss of cerebellar input, which then alters the overall CNS properties. We feel the key in this integration/coordination is the cerebellum's possible role in the dual processing of the hemispheres (see following paragraphs). In this speculation we hope we are not exaggerating this organ's role.

Supporting evidence for cerebellar input regarding the integration and coordination of the two cerebral hemispheres (and their associated limbic and motor subsystems) is seen in the data of Klein and Armitage (1979), who claim that the left and right cerebral hemispheres operate cyclically in their efficiency of operation, on a 90- to 100-minute basis, exactly 180 degrees out of synchrony with each other. Thus, hemispheric collaboration is orchestrated or coordinated in a highly regulated manner that appears to give the two hemispheres serial access on a guaranteed basis. It is possible, according to Klein and Armitage, also to understand why REM dreams differ from non-REM dreams (and may continue as an electrophysiological pattern throughout the daytime as well): they represent the activities of the dominant and nondominant hemispheres, respectively (Corhalles, 1978, pp. 1326–1327).

The work of Klein and Armitage is not, however, so straightforward as it seems. Kripke, Mullaney, and Fleck (1984) failed to replicate Klein and Armitage's experiment and seriously question the validity of the conclusion that the hemispheres alternate. (For those interested in the details of this research on brain rhythms we recommend Kripke, 1982, and Kripke et al., 1984.) Our conclusion that the cerebellum may make a contribution here to the dual processing of the hemispheres is, of course, based only on that part of Klein and Armitage's work which *has* been replicated. In addition, our speculation follows from our sense of how the cerebellum works, which we shall shortly discuss in more detail, as well as from the observation that there is usually a fit between the particular hemisphere that is activated and the cognitive task to be performed (see Gur and Gur, 1980). The cerebellum seems to carry out the cortex instructions by maintaining a running dialogue with the cortex by way of cerebellar "comments"; we wonder whether the cerebellum might not be capable of including in its commentary on how a particular activity might be accomplished some judgment as to which hemisphere might better be used or how the hemispheres might be used in what overall patterning. For example, a doctor in an emergency room needs his left (dominant) hemisphere but not his limbic system, during a crisis. Only later on, after the emergency, when strong affects will not influence the doctor's use of his medical knowledge, is the limbic system likely to be involved in the appraisal of what just happened. In contrast, a musician or psychoanalyst would be helpless without the combined use of both hemispheres, and some activities (such as play) might best proceed with the right hemisphere alone (so as to avoid inhibitions).

Aschoff (1981a) describes the short-term activity rhythms of the type observed and measured by Klein and Armitage (1979), indicating that they are known to exist in a number of different vertebrate species, where they appear to contribute adaptive advantage to the group (for example, by assuring that all members of the group are not asleep at the same time). Their discovery in man was predicted by Kleitman (1963), who proposed that the basic rest–activity cycle (BRAC) continues in adult humans during the daytime while at night it demonstrates itself in the phenomenon of REM and non-REM sleep (p. 492). Friedman and Fisher (1967) were also pioneers in this area of research. The mechanisms of such neurologic periodicity are unknown but currently under active investigation, and the interested reader might wish to consult Aschoff (1981a,b) for a review of this complex subject.

In chapter 2 we quote Galin (1974) as suggesting that the hemispheres might be integrated by "taking turns" and that this alternation might involve one hemispheres inhibiting the other (p. 575). We think Galin could be right, but we want to add that such inhibition might occur either directly or indirectly, through a cortical instruction from the cerebellum. We tend to favor the latter possibility but have insufficient data at the present time to come to any definite conclusion. There is, however, at least one precedent for the idea of inhibition by way of the cerebellar stimulation and inhibition of antagonistic muscle groups, since the cerebellum carries out the motor purgatives of the cortex. (Of course, such reciprocal muscle stimulation and inhibition are also the consequence of reflex arcs organized at levels below the cerebellum; thus, the mechanisms involved in such coordinated activities of the motor apparatus are certainly multiple).

Gur and Gur (1980) review what is known about the relationship between cognitive-task and hemispheric activation, as measured by such variables as EEG, regional blood flow, and the direction of cognitive eye movements. There apparently is some tendency to use a particular hemisphere on the basis of its preferred cognitive task; however, little is apparently reported on how the two hemispheres may jointly tackle a task in a complex, coordinated manner (see Moscovich, 1979). Of particular interest is whether or how such complex patterns of hemispheric activity over time may reflect complex information-processing strategies. Moreover, it seems likely that such strategies would be learned and subject to learned revision (as, for example, in a psychoanalysis).

Kinsbourne (1980) confirms that relatively little is known about what determines complex patterns of hemispheric activation, but he proposes two possible mechanisms: There is evidence that the brain stem itself has some asymmetry, which may contribute to the asymmetry or specialization tendencies of the hemispheres in problem solving. There is also some evidence, but no description as yet, of a "brain stem switcher mechanism" (p. 180), the details of which appear to be a major missing link in our understanding of how the hemispheres collaborate. In the same vein, we need to know more about brain-stem "gating."

Moscovich (1979) points out a number of complexities regarding information processing that are important to consider in our efforts to understand how the hemispheres accomplish their tasks. He begins by integrating cognitive psychology approaches with brain-localization research, reminding us that although one can take the position that in psychology it is enough to know that something is happening, nevertheless by also knowing where and how in the nervous system these functions are carried out, one can begin to "elucidate further the nature of [these] . . . processes themselves" (p. 382). To accomplish this synthesis, Moscovich starts with the currently accepted memory model of information processing: stimuli are the input to a sensory store (sensory phase of memory) that selectively encodes this information and inputs it into short-term memory. The output of the latter is then input for long-term memory (see either our earlier discussion or p. 399 in Moscovich for details).

Moscovich then asks where the locus is for brain asymmetry. After reviewing extensive neuropsychiatric data he concludes that "the peripheral processes concerned with extraction of [data on the] physical features of a stimulus are actually common to both hemispheres" (p. 388); that is, the hemispheres are not different (asymmetric) in their ability to accomplish a feature analysis (color, contrast, loudness, pitch, etc.), but the "hemispheric asymmetries in information processing emerge only at a higher level of analysis in which relational or categorical features are represented") p.411). This coincides with the stage during which data representing the encoding of the sensory input become an input for short-term memory. There is a great deal of anatomical evidence that this stage involves processors at the visual and auditory primary-association cortical level.

Moscovich next tackles the problem of hemispheric collaboration, which for him is linked to localization and laterlization. He considers a number of hypotheses, such as the familiar ones of sequential processing versus analytic/holisitc processing, but ends up concluding: "Whether the operation of the hemispheres is best described by a single principle, or a multiplicity of them, is a problem that is not likely to be resolved soon. What is not an issue is that two fundamentally different systems exist that are capable of processing information somewhat independently of each other" (p. 417).

Finally, however, he addresses himself to the possibility of dual processing, which follows logically from his conclusion that the two hemispheres often start out with the same data at a sensory level. The question then becomes how dual processing might be accomplished. The reader is referred to his discussion (pp. 417–422) for details, but the most salient issue appears to be how one might imagine the cooperation of the two dissimilar hemispheric systems. Whether there are two different systems or a single system for mental representations is a matter of controversy. If one assumes two representational systems, how, where, and whether data might be exchanged between them are also matters of controversy. Moscovich feels that if the question is left at the philosophical level and depends on proving "that words and images are (or are not) phenomenological manifestations of similar abstract propositions" (p. 421), then the debate will continue for a long time. However, considering that each hemisphere most often processes the kind of information for which it is specialized, Moscovich suggests a collaborative mechanism as follows: "If information in the left and right hemispheres were represented in fundamentally different forms, these differences might preclude exchange of information between the hemispheres unless a third, abstract code existed that encompassed both of them" (p. 421).

How might such an "abstract code" work?[1] First of all, it would seem to be necessary only after the primary sensory-modality-processing level has occurred, inasmuch as before this time and level are reached the two hemispheres contain, through their parallel processing, exactly the same information and no bridging would be necessary. However, after this level the data would begin to diverge, and the associative cortices for each of the primary sensory modalities would then need a mechanism to allow dual processing. Bridging might occur in a number of ways in the adult; or in the child once sufficient myelinization occurs, a number of key commissures (e.g., the hippocampal commissure or corpus callosum) will connect the hemispheres. Even without commissures however, the bridging could still occur by having information arrive at the contralateral cerebellum from each hemisphere at the same time. This information could then pass back to the cortical area from which it came but only after being modified so as to take into account the information/input simultaneously in the cerebellum from the other hemisphere. This modification or mutual influence of hemispheric inputs to the cerebellum from each hemisphere would need to involve a "code for translation," or "code" could stand for any mechanism that would transduce the data, that is, allow it to pass from one cerebellar hemisphere to the other.

To illustrate this speculative cerebellar role, we are reminded of the story of a researcher working on an archaeological project that seemed to defy solution because his data suggested that a particular structure or group of structures under excavation seemed to have originated several hundred years before the accepted date for the site. While being held over in an airport because of some delay, he happened to meet a fellow passenger who, by chance, was also involved in work on the same subject; from their discussion the researcher learned that the date of the site had just been corrected by exactly the number of years that his data suggested!

It is possible that the cerebellum may constitute such an airport or meeting place and that the differing information coming into and leaving it from each hemisphere (and other areas?) may capitalize on the simultaneity of these "events" and an information exchange might occur, by whatever means the cerebellar "translator" has at its disposal.

---

[1]In our opinion this question deserves a separate paper that would include a review of the work that bridges neuropsychiatry, psychology, and artificial intelligence theory. To date, the most lucid exposition of such an interdisciplinary sort is the work of Kent (1981).

CEREBELLAR ANATOMY AND PHYSIOLOGY

For a detailed review of cerebellar anatomy and physiology the reader is referred to Eccles (1979), Frick (1982), Itoh (1984), Palay and Chan-Palay (1974, 1982), and Shepherd (1979). A brief description of cerebellar–cerebral connections and relatedness may, however, help clarify what we are suggesting regarding the major learning role for cerebellar input. The cerebrum and cerebellum connect with each other not directly but rather by way of brain-stem nuclei. The cortex sends efferents from two types of pyramidal cells in the motor area: large and small cells. The large cells send fiber tracts to the ventrolateral nucleus of the thalamus (VL), the red nucleus (RN), and the pontine nucleus (PN). Fibers descending from the small pyramidal cells enter these nuclei but also send some fibers, through the red nucleus, to the inferior olivary nucleus (IO). Fibers from all these brain-stem nuclei now cross the midline and enter tracts running toward the cerebellum. Those from the IO nucleus are unique: so-called climbing fibers run directly to impinge on the Purkinje cells of the cerebellar cortex, providing the only excitatory (direct) input that the Purkinje cells receive. The fibers from the PN and RN send mossy fibers to synapse with the so-called granule cells near, but do not make contact with, the Purkinje cells. The granule cells give off parallel fibers in abundance that do synapse with the Purkinje cells, providing both excitatory and inhibitory input in a dispersed but powerful system. Eccles (1982) feels that the climbing fiber input from the inferior olivary nucleus plays a key role in cerebellar learning (p. 612).

We are, therefore, describing Purkinje two major types of input fibers: climbing fibers originating mainly in the inferior olivary complex and mossy fibers with more widespread origins. The inferior olive, in turn, receives input from "all three levels of the brain: spinal cord, brain stem, and cerebral cortex" (Shepherd, 1979, p. 216). In addition, the locus ceruleus also sends monoamine fibers to the cerebellum.

The brain is usually described as containing 100 billion nerve cells, the cerebellar hemispheres together containing 40 billion granule cells (Shepherd, 1979). Obviously the cerebellum is usually undercounted in importance. This omission is interesting considering that the cerebellar cortex contains seven times the number of output neurons as the retina and more than 100 times the number of similarly functioning cells as the olfactory cortex. "From these numbers alone one would anticipate that the cerebellum must provide a dominant input to the brain" (p. 221). Remarkably, however, there still is no consensus as to what the cerebellum does, other than that it helps initiate and maintain some types of movement and posture (Shepherd, 1979; Thach, Perry, and Schieber 1982).

The arrangement of cells within the cerebellum actually creates an enormously interesting and novel situation within the brain: "sustained activity in response to an input is not possible within the cerebellar circuits" (Shepherd, 1979, p. 235). That is, there are no pathways for reverberating activity! "The extreme, high frequency burst elicited by the climbing fiber, followed by suppression, reflects the specialization of this input for eliciting clearly detectable Purkinje's cell response no matter what the background discharge" (p. 236). According to Shepherd, the cerebellum is seen as a "massive accessory processing apparatus superimposed on the input–output relations of the nuclear cells with the rest of the brain" (p. 237), with unique abilities to identify changes occurring in some background pattern of nervous activity.

It is interesting to compare the cerebellum and the retina, the latter providing "graded potentials" in contrast to the cerebellar processing in the "frequency domain" (p. 243). On one hand, this means that the cerebellum operates more in a digital mode whereas the retina operates more in the analogue mode (p. 243). On the other hand, both the retina and the cerebellum seem to function by processing information with transient responses in the form of "perturbations of ongoing activity." Such a system apparently allows these systems to function with a high degree of accuracy (compared with what would be the case for information transfer by transmission by excitatory responses against little or no background [p. 243]. Clearly, there are functional reasons for the existence of such unique computational structures as the retina and the cerebellum. One speculation is that the digital structure of the cerebellum (also see Marr's, 1969, work in this regard) provides the cerebellum with some unique learning capacities.

The indirect input to the cerebellum from the cortex is very fast. The cerebrum cannot begin instituting any action without the cerebellum's immediately knowing about it. Although the cerebrum unquestionably is the "command center," all the instructions the cortex fires for motor activity to the machinery of the spinal cord are simultaneously delivered to the "computational machinery" of the cerebellar cortex (Eccles, 1979, pp. 6–7). In humans it takes 1/50th of a second for a complete loop of the circuit to be made from cortex to cerebellum and back to cortex via critical brainstem nuclei (p.9). This is a closed loop. In the case of the cerebellar hemispheres, however, the loop is open, and the cerebellar hemispheres, which make up 90% of the organ, receive most of their cerebral input from extensive areas of the cortex, such as the motor-association cortex and the supplementary-motor cortex (p. 10). When orders for movement are planned or executed, the cerebellum receives the initial instructions and feedback regarding the initial results of all movement, provided by feedback loops from sensory receptors in the periphery. The cerebellar cortex then computes a "comment" that is returned to the same area of the motor cortex that originated the action order (p. 12). According to Eccles, *"the cerebellum's role is largely anticipatory, based upon learning and previous experience, and also upon preliminary, highly digested sensory information transmitted from some of the association cortex"* (p. 14) and less refined inputs from the spinal cord. We can summarize by saying that the cortex acts as the commander of the ship, the cerebellum is comparable to the executive officer, and the motor apparatus represents the troops. The commander and executive officer need to send and receive messages to each other continuously about the movements of the troops if the system is to function satisfactorily.

The cerebellar input can be blocked by various brain-stem nuclei that serve as critical connections in the indirect loops that provide for connection between the cerebellum and the cerebral cortex and higher centers. We considered this possibility when we discussed the work of Winson (1985) on the hippocampus, where gating is also critical. When this cerebellar input is "gated" out, and not available, the higher centers would be expected to function without the benefit of the cerebellum's "comments." The cerebellum will further "suffer" because it no longer gets feedback about the effects of its "comments" on the action plans of the cortex and higher centers. In short, the entire system will operate without the richness of integration and coordination that the cerebellar input can provide.

But of what does this cerebellar input consist? Included in the commentary to the cortex and higher centers might be information about how the cortex can best accomplish its tasks by patterns of reciprocal inhibition and or facilitation of the two cerebral hemispheres. That is, the cerebellum not only has ideas about how to guide successful muscle activity by knowing how and when to stimulate and when to inhibit which (antagonistic) muscle groups, but it may also store memories about the experience of the individual in using its two cerebral hemispheres, either in isolation or in combination, and might be able to make specific suggestions about these dual-processing strategies. In addition, the cerebellum may provide some other special "comments" to the cortex and other systems (Massion and Sasaki, 1979) that may become important for memory and learning. We shall discuss this after some brief comments on the work of Itoh on brain mechanisms and cerebellar organization.

THE CEREBELLUM AND THE STORING AND RESTORING OF PROGRAMMATIC MEMORIES

The cerebellar role has been investigated by many authors, probably too many to summarize briefly. Itoh (1970, 1976, 1981, 1984a, b, c) has followed up on his work with Eccles and Szentagothai (1967) as well as that of Marr (1969) and developed what he considers a paradigmatic case, the VOR reflex-arc hypothesis, which exemplifies systems-control functions within the nervous system. The anatomy of this organ has been studied in extraordinary detail and beauty by the Palays (1982). Those interested in this subject should also consult Eccles (1979) for an elegant but simple description of cerebellar functional anatomy.

Body maps appear in the cerebellum, and the cells controlling different parts of the body "are arranged within the cerebellar nuclei in a somatotrophia" (Thach, Perry, and Schieber, 1982, p. 440).

In addition to the control of integration between the two cerebral hemispheres, a second possible role for *the cerebellum might involve the storing and "restoring" of critical but nonretrievable information of a programmatic nature.* We believe these activities would constitute a major contribution to learning in general. These functions are also accomplished by other brain subsystems, such as the limbic system; of course, the design of redundancy in the nervous system is clearly one of its central characteristics.

Any electrical system can be expected to have "electrical noise," which will alter the memories or programs that are stored within it. The brain is no exception to this rule, whatever its systems to dampen or eliminate such factors. One way the brain protects against noise is by building redundancy into the system, in general. However, a more specific mechanism of a speculative sort is the restoration of programmatic memories from one substructure to another substructure of the brain to help overcome problems in retrieving memories that have suffered changes in the formatting required for retrieval. By analogy, when this occurs in a document we are working on with a computer, we simply restore the memory in a hard disc by inputting a floppy disc backup of the original. Something similar may occur to explain certain kinds of forgetting. In these cases, the psychoanalytic interpretation functions as a floppy disc backup. Since the differences between the backup and the unretrievable original are negligible, nothing in the way of change is noticed; and once restored, the memory crystallizes a process of further memory reconstruction. This may be the reason why "priming the pump" is at times needed in analysis.[2]

[2]Priming is further defined in relation to procedural (versus declarative) memory systems in chapter 6, footnote 2. The topic of priming is important for psychoanalysis because of the role of the analyst's comments in inviting the analysand's memory (regarding those kinds of experiences that seem to require a perceptual event for their recall); it is also important for neuroscientists who wish to understand the phenomenon of remembering. Put somewhat differently, the

In regard to electrical noise, we are well aware of the insights of Pinneo (1966) on the importance so far as information processing is concerned of both "phasic" and "tonic" activity within neurons. In mentioning "noise," we are considering here something other than these kinds of nervous activities, namely, the vulnerability of specific CNS storage mechanisms (see also Palay and Chan-Palay, 1974, particularly their discussion of the role of glial cells).

There are a number of places from which such backup may come: the hippocampus, various associative areas of the cortex, the diencephalon, the gyrus cinguli, and the cerebellum. The cerebellum is likely to be important in long-term memory storage; a great deal of evidence indicates that one of its major functions is to back up the cortex regarding motor activities. Thus the cerebellum might contain, in motor language, duplicate action plans (memories) of virtually all of our activities, both executed and fantasized. And, as Piaget has concluded (Piaget and Inhelder, 1969), it is from the motor or action viewpoint that we begin to understand much of our world. Furthermore, the cerebellum is present as a major connection between the cerebral hemispheres from the beginning and long before the major commissures linking the hemispheres complete their myelinization and function fully. On this basis alone, the cerebellum would seem to have an advantage in establishing its dominance over the hemispheres (much as, on the basis of the left-over-right timing of these connections, the right cerebral hemisphere assumes a dominant connection with the limbic system, and the language system or left hemisphere assumes control over the motor apparatus).

Following up on his pioneering book with Eccles and Szentagothair, (Eccles, Itoh, and Szentagothair, 1967), Itoh (1984a) has continued to explore the cerebellum as a computer. Although he observes that we still have "no clear vision of how the cerebellum achieves its function" (p. vii), he nevertheless concludes that this organ appears to endow a motor or autonomic system with three major capabilities—coordination, orthometria, and compensation—"just as a modern computer endows an engineering system with multivariate, predictive, and adaptive-learning control capabilities" (p. x). Itoh even feels that cerebellar plasticity may have a different molecular basis from the plasticity that has been discovered elsewhere in the CNS, for example, in the hippocampus (p. 121) or in aplysia ganglia research. He is also trying to identify exactly whether the cerebellum is making "comments" to the cortex itself or is merely acting in

---

existence of priming is one more piece of evidence suggesting that human memory represents the activity of a variety of systems or modules, each with its own qualities or mechanisms for encoding and retrieval (Tulving and Schacter, 1990).

concert with some extracerebellar system (p. 130). Basically, he concludes that his work on the special reflexes of the cerebellum (such as the VOR reflex arc, which is responsible for maintaining the focus of any retinal image, whatever the movement of the observer or the environment) is entirely in keeping with his view, shared with Eccles and Szentágothai, that the cerebellum is a learning-control machine par excellence (p. 465).

We find the work of Itoh and his associates compatible with our hypothesis; specifically, we cannot help but wonder to what extent learning depends on exposing experience to that part of the nervous system whose plasticity is most likely to capture and benefit it. In this sense there is reason to believe that the cerebellum keeps much more "in focus" than retinal images.

DISCUSSION

The complexity of the preceding sections may make it difficult for the reader to have a capsule view of how we are answering the questions with which we started, namely, how learning is accomplished, how this is related to the plasticity of the brain, and, still more specifically, how the two cerebral hemispheres collaborate in the learning process. We shall present our overview of the data from three points of view: meaning, relevance, and application.

On the basis of the work presented, we are suggesting the following conclusions regarding plasticity: (1) the brain can change at each level of organization, at multiple localizations, and at any time within the life cycle; (2) these changes involve anatomical, physiological, psychological, and chemical changes, depending primarily on one's point of view; (3) they are enhanced by emotionally meaningful input, as well as by chemicals or so-called biological approaches; and (4) these changes represent species-specific behavior and have "releasers," as well as some genetically controlled mechanism for the timing of their expression.

Regarding learning, we conclude that (1) learning is possible because of the brain's plasticity; (2) learning involves tapping all of the plastic potential of the brain; (3) learning can be blocked if any level or any anatomical or functional system of the brain is disturbed, for any of a variety of reasons; (4) learning may involve simple responses, such as habituation and sensitization, or more complex, discriminative, or associative kinds of responses, depending on whether or not the so-called reward system of the brain is involved in facilitating the consolidation of short- to long-term memory storage of whatever is being learned; (5) learning involves some "learning readiness" that requires priming; (6) this priming involves the activation of a number of major subsystems of the brain adapted to associative learning, such as the limbic system, the cerebral cortex of each side of the brain, and the vestibulocerebellar system; and (7) the two hemispheres can accomplish their learning together or in tandem.

Regarding the coordination of the two hemispheres, we conclude that (1) the best evidence suggests mechanisms for the coordiation of their collaboration even before the major commissures are myelinated; (2) whether or not they alternate in their activity, they certainly go through 90- to 100-minute cycles of peak performance; (3) whatever the coordination mechanism, it seems to involve assuring a match between the particular hemisphere chosen for a task and the specific cognitive task being tackled; (4) a number of brain subsystems appear to be involved, and these include the cortex, the limbic system, the brain stem, and the vestibulocerebellar system; (5) because of its unique availability early in life, as well as its anatomy, connectedness, and special capabilities, the vestibulocerebellar system appears to contribute to learning and memory; (6) one possible cerebellar contribution is that in carrying out the cortex commands to accomplish a given task the cerebellum first specifies which hemisphere we use in which order, thus creating the ongoing pattern of dual processing; and (7) the cerebellum may also be involved in directly bridging the hemispheres by means of a "code" that allows the output of one hemisphere to translate into an input of the other.

During the course of a psychoanalysis a number of mechanisms will be tapped that seem conducive to learning. Although we cannot be exhaustive, we note the following: (1) personal and emotional input alerts the nervous system to ready itself for input (see chapter 1); (2) timely soothing activities of the analyst probably reduce critical brain stem gating in the analysand so that the hippocampal/limbic and vestibulocerebellar systems are functional and contributory to the overall information processing of the brain; (3) the analysand identifies new information processing methods based on a modeling after the analyst's methods (including, more specifically, how and when the analyst uses which hemisphere) and a judgment as to the adaptiveness of the viewpoints and assumptions implicit in the analyst's modus operandi; (4) some critical bridging occurs between the hemispheres that has the function of undoing repression and disavowal (see Basch, 1983; Galin, 1974; chapter 2, this volume), (5) some analytic interpretations will result in the restoration of information (stored but inaccessible) by means of its replacement with "earlier versions"; (6) some of these "earlier versions" will come from the cerebellum under the stimulation of the analyst's interpretations and reconstructions; (7) in the analytic process the analyst and analysand will make use of their vestibulocerebellar systems to provide and communicate empathy; and (8) there is the possibility that some critical memories will be recovered by the assumption of postures or the reenactment of subtle action patterns that were part of the experience around which those memories were originally organized.

As for the relevance of the work on bridging neurology and psychoanalysis, although one can clearly be an exceptional psychoanalyst without knowing anything about the brain, it seems likely that psychoanalysis, in both its methodology and its theory, would benefit from the awareness and utilization of the explosion of knowledge in the neurosciences in general. It seems especially likely that if analysts do not recognize the need to bridge the disciplines of neuroscience and psychoanalysis, other specialists will do so and will then take credit for the insights and progress so obtained. The early progress of Freud, Hartmann, and many others is attributable, in part, to the fact that they were not afraid to take advantage of their knowledge of the brain, although they clearly extended this knowledge in powerful ways nonbiologically.

In dealing with the question of relevance, we wish to remind the reader that learning disabilities are probably not uncommon in analysands. In fact, each of the analysands described in the clinical vignettes suffered from some kind of learning block. When we discover such a learning deficit, how are we to understand it? If it resolves during analysis, what is implied about etiology? What if it responds to medication? Clearly, a multidisciplinary approach seems more logical in some cases than remaining exclusively within our area of expertise and not asking for consultations, when appropriate, from experts in learning problems. It seems to us that psychoanalytically informed neurologists and neurologically informed analysts are in a privileged position to make observations that would help with both the solution of specific clinical problems and our general understanding of brain mechanisms.

More specifically, we feel that emotional problems might be considered simultaneously from a number of points of view in addition to the traditional psychoanalytic one. Two kinds of studies immediately suggest themselves on the basis of our survey and, in particular, our review of the possible coordinating and integrating role played by the vestibulocerebellar system in learning. The first kind of study would collect the data on children and adults with documented disturbances of gaze and determine if these subjects showed any particular clustering of specific kinds of learning or information processing difficulties (Perlo, 1985, personal communication). The second kind of study would involve testing extraocular eye movements of selected analysands with learning blocks to determine whether, and what kinds of, gaze disturbances are present. The results of such a study might help not only to establish some simple screening tests to diagnose and analyze learning blocks but also to improve our understanding of the ways in which people learn.

SUMMARY AND CONCLUSIONS

Our interest in the effectiveness of transference interpretations in opening the gates of memory and overcoming learning thresholds led us to an examination of how the mind/brain integrates its hemispheric activities, suggesting the possibility that psychoanalytic interpretations may serve as critical bridges between the two cerebral hemispheres, overcoming repression and disavowal, which appear to us to be left-to-right and right-to-left interhemispheric communication blocks, respectively.

We subsequently went on to a more detailed investigation of how the hemispheres are coordinated in their activities and to a review of some of the current research on brain plasticity and learning bearing on this question. In the process we learned that, unlike the brains of lower animals, which appear to be relatively hard-wired, the brains of humans are enormously plastic at almost every level of their hierarchical and "triune" organization. Thus, in tentative answer to the question of how nature and nurture interact, we might state that at each level of organization within the brain it is possbile to specify what the inborn givens are and how specific environmental experience can begin to tap the plastic property of each level of the nerve structure. There is no reason to assume that psychoanalysis exerts its effects at only one level within the brain or that its impact on the brain is in any way fundamentally different from so-called organic or biological approaches to mental illness.

We reviewed some of the data from neurology that bears on the ability of the genome to respond to the environment adaptively and express parts of our inheritance previously hidden. We then described various levels of change or adaptation, including some of the latest work in search of the engram. We discussed how certain behavioral circuits can carry short-term memory and how these can be put into more permanent, or long-term, storage by input from some of the brain's critical motivational machinery, such as the locus ceruleus. We also learned that at least three systems are critical for the kind of discriminative learning that we associate with human behavior at its most complex level: the system of the right hemisphere, with its preferential attachments to the limbic system; the left hemisphere, with its motor system dominance; and the vestibulocerebellar system. We also learned that critical brain-stem nuclei can either glue together or unglue these major subsystems of the brain by electronic "gating."

Along the way we considered the interesting possibility of viewing psychological problems from at least one other perspective—that of learning disabilities—besides the ones ordinarily applied in the clinical psychoanalytic situation. This departure fits well with the idea that (1) the brain is organized according to many principles, and (2) psychoanalysis can alter these organizations. Also, our earlier views on left-right brain mechanisms and repression/ disavowal phenomena certainly fit within this perspective. So, apparently, does a large part of the data from our review of learning and brain plasticity. We seem to be living at a critical time in which the convergence of brain sciences and behavioral specialties is leading to insights into fundamental mechanisms of nervous system defense, organization, and control. But we do not understand things so well that we can afford to ignore either neurological or psychoanalytic insights into learning blocks.

Finally, to return to our original question of understanding how the cerebral hemispheres coordinate their activities with each other, we considered possible roles for the cerebellum, which seems to have been relatively neglected in the study of learning, memory, organization, and coordination. In our consideration, we relied on some of Freud's earliest insights about empathic observation, which suggest a dual processing of the hemispheres. We concluded that the cerebellum possibly makes at least three major contributions or inputs to the overall functioning of the nervous system, each of which could be expected to have a profound role in learning.

The first contribution stems from the possibility that the bridging between the cerebral hemispheres involves, to some extent, a bridging of the cerebellar hemispheres, with the latter based on the translation of each hemispheric output into a code that becomes the input for the other cerebellar hemisphere. In making this suggestion, we are following up on some insights of Moscovich regarding dual processing.

The second contribution involves the possibility that cerebellar input may be critical in determining which cerebral hemisphere to use at a given time, or whether to use both. Obviously, there are also unknown brain-stem mechanisms that may critically affect hemispheric cooperation by means of their gating effects, but we nevertheless feel that the cerebellar input may be a part of this complex interhemispheric coordination process. There is also evidence that under some circumstance the two hemispheres cycle in their activity, thus giving them serial access to experience for those times when they are not in synchrony. This cycling, if it proves a reliable observation, may also reflect the direct or the indirect inhibitory effect of one cerebral hemisphere on the other.

The third cerebellar contribution would be the provision of backup for programmatic memories that cannot be retrieved because electrical events within the brain have caused alterations in the formatting required for retrieval. We stated a number of reasons that make this hypothesis attractive, but we admit it is at present speculative. It is, however, in keeping with one of the main features of the brain's organization, namely, its redundancy, which provides safety through backup.

We feel that these hypotheses are valuable for two reasons: on one hand they seem eminently testable. On the other hand, they begin to suggest how psychoanalysis is such an effective tool. Therapeutic interpretations have the capacity to provide bridges of extraordinary learning influence within the CNS. That is, they create a learning readiness, as, for example, one can imagine when a message reaches one of the patient's cortical hemispheres and results in a diminution of the "gating" at brain stem nuclear levels, secondary to cortical input of a disruptive sort. When the gating abates, the cerebellum, hippocampus, and other critical parts of the brain's controlling systems may be brought into service. In the case of the cerebellum there ensues a volley of communication between the cortex and the cerebellum that benefits each and greatly increases the quantity of past (learned) experience brought to bear on the current situation. It is also likely that the analysand learns as much from identifying with the analyst's methodology and mode of being as from the specific content of the analyst's interpretations.

In summary, at the highest level of nervous system organization, a critical learning threshold seems to be crossed when three parts of the CNS are brought together simultaneously; the right cortex/limbic system constitutes one part, the left (dominant) hemisphere/motor system is a second, and the vestibulocerebellar system is the third. Psychoanalysis facilitates the integration of these three components into a cohesive system with extraordinary learning potential.

Finally, it is our opinion that neurologically informed psychoanalysts and analytically informed neurologists, working together (as well as in consultation with other disciplines), are in a privileged position to take advantage of the convergence of their respective fields and clarify the basic workings of the brain.

4

The Prefrontal Cortex and Neural Control

The Brain's Systems for Judgment, Insight, and Selective Attention

PRÉCIS

Freud (1900) relates the famous dream of Maury, lying ill in his bed, with his mother nearby. He is surrounded by scenes of murder from the Reign of Terror, including Robespierre's tribunal with Marat, Fouquier-Tinville, and others. Maury is questioned, condemned, and led to the place of execution, where he is bound and guillotined. He awakens from his dream with a pain in his neck and the feeling that, his head is separated from his body. What happened during his sleep was that in reality, the top of his bed had fallen on his neck. What happened in Maury's unconscious is less certain, but according to Freud (1900, pp. 26-27, 497), Maury has made use of both current sensory input and unconscious dream thoughts. The latter is seen in such evidence as the choice of setting, which shows the dreamer's ambitions in having surrounded himself with famous figures from the French Revolution, and "secondary elaboration," indicating that this kind of fantasy was familiar to Maury; partly explaining how the large quantity of detail becomes compressed into Maury's dream imagery within the short space of time allotted after the bed falls on his neck. Clearly, even allowing for some prepackaging of the dream content, the brain is capable of the extremely rapid integration of unconscious wishes and external (sensory) realities within complex scenarios.

Chapter 4 attempts to understand how such out-of-awareness integrative activity is coordinated. More specifically, it asks how native lan guage and the language of the brain might be related to each other. To this end, the work of Niwa (1989) at Tokyo University on the brain's operating system, and the research of Tsunoda (1987) at Tokyo Medical Dental College on the Japanese brain, are correlated with the investigations of Kent (1981), Ingvar (1987), Itoh (1988), and others on the subject of neural control. The prefrontal cortex is shown to play a decisive role in this mental coordination function.

The work of Tsunoda (1987) is bound to surprise and excite, for he has studied how the brain lateralization characteristics of the Japanese, for various dichotic listening tasks, are different from those of westerners in general, but not different if the westerners are fluent in the Japanese language. In other words, there is some data suggesting that fluency in the Japanese language can invite those who use it to listen to certain sounds, for example, the sound of human laughter or crying, with different brain structures than one would ordinarily expect to be used. Some possible implications of this are discussed.

This chapter considers psychiatric, neurophysiological, information processing, and psychoanalytic aspects of adaptation, understood under the rubric of neural control. "Neural control" is another name for the complex, adaptive, internal decision making of the brain. The rules for neural control have traditionally been of interest to psychoanalysis, but have seldom been approached systematically. Consequently, there is little in the psychoanalytic literature regarding what such control (i.e., its mechanisms) actually consists of or about its possible anatomical-physiological basis. In chapters 1 through 3 I have contended that neuroscientific and psychoanalytic insights can be usefully employed together. In the process, I have delineated a plausible neurophysiological basis for such familiar analytic phenomena as transference interpretations, psychological defense(s), learning, dreams, and early psychological development. In the following, I describe and define aspects of insight, judgment, and selective attention, which I believe express the brain's system for "feedforward neural control" (Kent, 1981, p. 194).

The chapter will necessarily contain some new technical information and may require some effort on the part of those unfamiliar with the subject. To simplify matters, whenever possible, new vocabulary will be defined as it is introduced. For example, the engineering expression "feedforward" (see Kent, 1981, pp. 194-203) needs to be defined before I continue further, as do the terms selective attention and insight. Feedforward regulation is in contrast to systems characterized by feedback (see Figure 1A). That is, systems in which there is a control loop describing how specific output becomes input, and therefore controls, modifies, or refines, future output of the system as a whole. Figure 1B shows such a feedforward system. In this system a feedforward controller operates

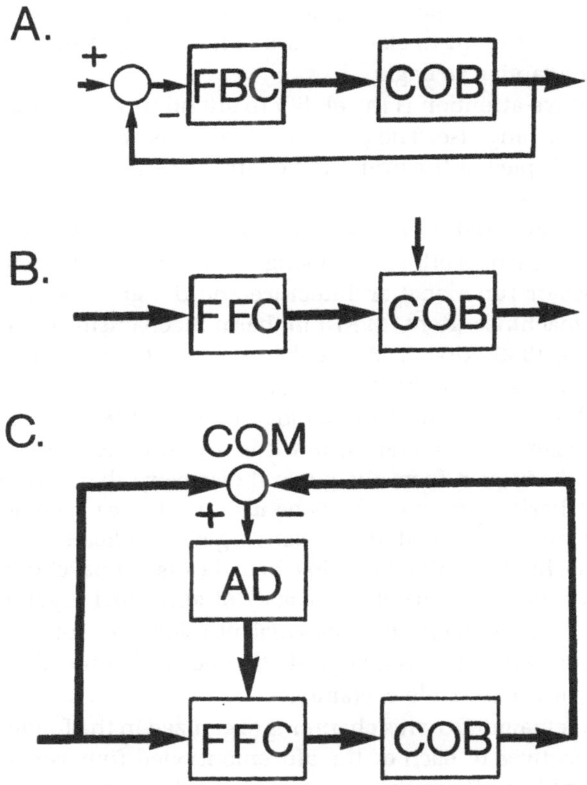

Figure 1. Block Diagrams for Control System. (From Itoh, 1988). FBC = feedback controller; COB = control object; FFC = feed forward controller; COM = comparator; AD = adjuster.

without benefit of any ongoing feedback. In practice, within the brain and many biological systems, feedback and feedforward systems are sometimes employed in parallel (i.e., simultaneously). Such a parallel system is diagrammed in Figure 1C.

Insight in psychoanalysis usually refers to understanding that occurs along with emotional growth and the resolution of pathogenic content and conflict. It is thus differentiated from cognitive understanding (see Moore and Fine, 1968, p. 55). Here I am employing the common sense, nontechnical meaning of having a keen sense of understanding. By judgment I mean the sense of being able to make decisions in some intelligent manner. From my perspective, these two terms blend into each other because they both seem to depend on the ability to anticipate future events and optimize conditions. In a sense, this entire chapter is an attempt to clarify possible meanings of these terms, in the context of adaptive struggle.

Selective attention is the ability to attend to something to the exclusion of everything else. The prefrontal cortex is critical for this function, since it is this part of the brain that controls the arousal of the rest of the brain other than itself.

The prefrontal cortex has been the special focus of a number of converging research efforts involving several different disciplines: (a) the cybernetics or functional architecture and design of so-called brains of men and machines (Kent, 1981); (b) basic mechanisms of schizophrenia (Andreasen, 1989; Niwa, 1989); (c) basic cerebellar mechanisms of neural control (Itoh, 1988); and (d) the investigation of higher level brain activity using real-time scanning methodology (Ingvar, 1983b). The reasons for this convergence are complex, but it is my contention that this convergence results in part from the general discovery that certain, recurrent (common), basic physiological mechanisms underlie psychological events. In particular, I believe that the convergence indicates that prefrontal cortical mechanisms play a role in a broad range of psychopathology. We may not yet have a clearcut physiology of abnormal psychology (or, one could say, a psychology of abnormal physiology), but some emerging correlations between psychological and neuroscientific observations can be made with reasonable certainty.

The remainder of this chapter is organized in the following manner: The perspectives of each of the aforementioned four research domains are described briefly as they relate to the subject and to each other. But first I introduce a functional consideration of the prefrontal cortex by reviewing the clinical symptoms associated with damage to this part of the brain. Then I integrate the various views of prefrontal cortical function with an emphasis on basic mechanisms and the clinical implications of such knowledge for psychoanalysis.

SYMPTOMS OF PREFRONTAL DAMAGE

According to Andreasen (1989),

> The frontal system is the brain region that mediates a large number of volitional, intellectual, and social functions. . . . Lesions . . . to the prefrontal cortex [in man] lead to disorders of cognitive function (concreteness, impaired attention, difficulty abstracting or categorizing), diminished spontaneity of speech, decrease in voluntary motor behavior, decreased will and energy, a tendency to engage in

repetitious or perseverative behavior, difficulty in shifting response set, and abnormalities of affect and emotion (particularly apathy, indifference, shallowness, and "witzelsucht" [p. 103].

Personality and temperament are affected, and there is difficulty in forming problem-solving strategies (Kent, 1981, p. 208). Kent refers to Luria's research, which compares the eye movements of normal people with those of persons with prefrontal damage. Subjects were shown a picture and asked various questions about it. "Normal subjects show an orientation of gaze in each case that depends on the kind of question asked about the picture and which reflects attention to the relevant set of details. Prefrontal damage cases have fixed patterns of inspection of the picture that do not change with the kind of question asked" (p. 208-209).

Kolb and Winshaw (1980), in a particularly comprehensive review, discuss 9 major symptoms of frontal lobe damage: (1) aphasia; (2) impaired response inhibition; (3) poor voluntary gaze; (4) poor recency memory (i.e., recollection of the correct order of memories); (5) reduced corollary discharge (i.e., the failure of the prefrontal cortex to send the normal signals to the posterior sensory cortex, which usually signals that some action has been ordered, the result of which guarantees that when the action occurs the "perceptual world" will not appear to move); (6) poor movement programming; (7) impaired spatial orientation; (8) altered sexual behavior; and (9) reduced behavioral spontaneity. With regard to the category of social behavior and personality, they note the appearance of "pseudodepression" and "pseudopsychopathy" (p. 293). The sexual difficulty has been little studied, but includes masturbation in public (especially with orbital frontal lesions), alteration in level of libido, or hesitation to perform sexually unless led through the activity step by step (pp. 295).

In animals, according to Kent (1981), experimental lesions to the prefrontal cortex result in "deficits in problem-solving ability" that relate to "deficient performance of delayed response and alteration tasks" (p. 207). Response tasks require the animal to observe some reward (e.g., food placed in some location) but to wait some time period before it is allowed to retrieve the reward. In alteration tasks, the animal must make one response one time, but a different (opposite) response the next time, in order to obtain the reward. Kent states that these tasks "require some ability at representing the relationship between the environment and anticipated rewards for a period of time" (p. 208). Key here is that memory per se is not the problem. Reiser (1985) reports work from Goldman-Rakic et al. (1983) indicating that the prefrontal association cortex is necessary for infant monkeys to recognize the continued existence of objects despite their temporary disappearance. Their findings, as well as those of Brozoski et al. (1979) in Goldman-Rakic's laboratory, show the essential role of dopamine in the memory functions of the prefrontal cortex.

The higher level functions that become deviant as a result of injury to this part of the brain include many that are of interest to psychoanalysis. The prefrontal cortex "is among the last portions of the nervous system to fully develop, some pathways becoming functional only at about twenty years of age" (Kent, p. 209). In what follows, we will explore some of the research within neuropsychiatry that converges on this prefrontal region of the brain. In the process, we will develop hypotheses regarding how the prefrontal system contributes to the brain's regulation of judgment, insight, and attention, and these, in turn, will help us understand adaptation and its breakdown.

THE BRAINS OF MEN AND MACHINES

Kent's (1981) book is the best concise definition that we have of how the human brain is designed. Its organization is broken down into input control, output control, and an intervening control of the goal direction of processing. What is important is Kent's description of the "optimization of function by motivational feedforward" (p. 194). He begins with the idea that we need to understand the effects of "sets" upon perception. For the visual system, for example, the time required for the identification of an object in the visual field will be greatly reduced "if the object to be identified is one which is to be expected on the basis of our current situation" (p. 194). This is a "feedforward" system procedure that essentially enhances the dynamic tuning process (involving, in the visual system, eye movement and an attentional shift) by preselecting "the appropriate categories of objects for the initial trial at matching" (p. 194). Kent calls this "framing" within the field of artificial intelligence theory. Efficiency is gained, primarily, by matching the observed object to be identified with the most likely interpretations.

Kent next examines how one specifies the set of most likely objects. Clearly, this function is based on some knowledge base within the brain that codes for the present circumstances and itemizes what the previous association has been according to some probability scale (i.e., what is most likely, what is next most likely, and so forth). We are essentially talking about the effect of context on perception. When such perceptual feedforward is applied to the goal-directed system itself, the perceptual system is directed to pay attention to selective features of the world as they relate to some specific goal. According to Kent, "This kind of influence can be exerted at all levels of the input system and can direct the active process of adjusting the center of gaze at selected features of objects. On a finer scale, it can adjust the center of 'attention' within the area of the center of gaze . . . " (p. 194-195). So far as it concerns the perceptual process in general, Kent is essentially defining attention as that feedforward mechanism "which is active in preparing the system for maximum rapidity or selectivity of response to a set of features deemed to be important by some higher center that has access to goal system information" (p. 195).

Most significantly, "*this feedforward control of processes in the input and other systems is exerted by portions of the prefrontal cortex and is manifested throughout the brain as an inhibitory effect which can suppress what is irrelevant*" (p. 195; italics added), and we experience this mechanism subjectively as attention. Figure 2 presents the basic mechanism of control of attention. In actuality this system for attention is somewhat more complicated than as diagrammed, but the basic feature of the attentional system is that it contains a reflexive level (the orienting reflex); a more advanced level above it in which the prefrontal cortex makes command decisions regarding what stimuli or systems are most relevant from moment to moment; and a still more advanced level in which logical operations themselves may shift or affect attention (p. 196).

As part of its "sculpting out" of attentional control from the brain as a whole, the prefrontal cortex influences the frontal eyefields, thus considerably influencing receptive field organization (p. 197). Hence, what are called voluntary eye movements essentially reflect the basic prefrontal control mechanism that we are discussing. The prefrontal cortex, in other words, points our eyes at what is important for our attention and affects the lower motor neurons by means of its influence at the level of the motor

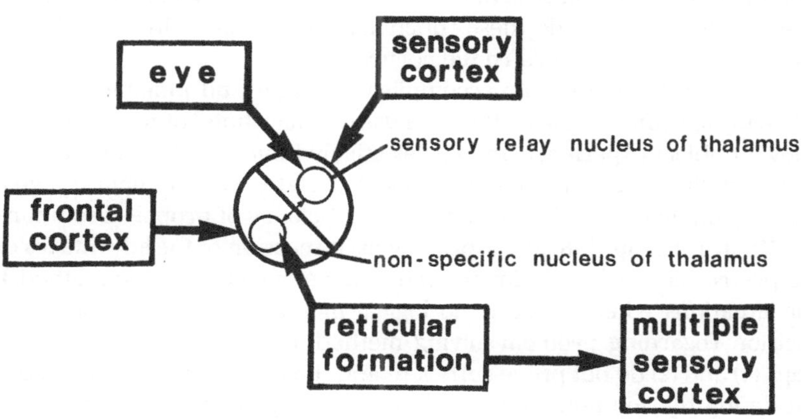

Figure 2. The basic mechanism of the control of attention. (From Kent, 1981, Fig. 7.4, p. 196). The prefrontal cortex can evoke specific activation of the reticular formation (RAS), in opposition to nonspecific activation of the RAS, at the level of the nonspecific nuclei of the thalamus.

supressor areas of the pons. The prefrontal cortex has enormous range and power directing the organism in relation to specific goals. The question remains, however, how the prefrontal cortex can make such decisions in a manner that is both insightful and expressive of judgment. The actions of the prefrontal cortex in regards to these functions,—of enormous significance to psychoanalysis,—will now be considered.

Kent describes a speculative model of how the brain functions in a complex problem-solving mode in humans, where the goal is not so much to find *the* solution as to find "good enough" solutions in complex environments. As previously noted, feedforward mechanisms can prepare "best guesses" based on learned associations, but "this feedforward preparatory process requires a supervisory system which can achieve a functional restructuring of all other systems" (p. 198). This anticipatory system, or what might be considered the highest level of the motivational system, seems also to be supplied by the prefrontal cortex, which has intimate connections with the limbic system and the lower brain stem motivational areas. Thus, it controls input, output, and higher cortical (motivational) processors as well. The question that we asked regarding how the prefrontal cortex carries out its function, then, breaks down to the question of how the functions of judgment and insight are accomplished, since these are what is required for the optimal functioning of the overall system. In other words, insight means being able to understand meaningful relationships between complex input and output variables, even when these relationships are not obvious or do not appear logical. Guessing correctly among a number of possibilities requires insight and judgment. But if my thinking here is generally correct, we still need to ask, in what concrete way do these functions become fact within the brain, specifically, within the prefrontal cortex?

The prefrontal cortex accomplishes insight and judgment by the following specific means: (1) It makes use of "snapshots" of motivationally relevant object experience and (2) selectively inhibits output operations, thus enabling it to (3) compare projected states with temporary goal representations to partially test out the sufficiency of proposed solutions (see Kent, 1981, pp. 198-204, especially his discussion of the advantages of the prefrontal cortex employing heuristic versus algorithmic methods). For situations where there is no logical method for making the initial decision regarding problem-solving method, or where aforementioned steps (1) and (2) do not prove sufficient and the temporary goal state does not define the solution, another step can be employed. In (4) the brain screens the current perceptual world for patterns that are similar to the desired temporary goal state, a task that appears trivial but is actually extraordinarily complex. Kent speculates that the same method used at the level of the brain's feature extractors (in the perceptual system) is then employed at the most abstract level within the brain, to make use of the brain's capacity to match patterns on the basis of similarity but, this time, the similarity is between hypercomplex features (see Kent, 1981, pp. 204-205). This search for similarities makes use of the extraction of further intermediates and of backtracking.

> This sort of approach has its origins in simple problems of spatial and temporal relationships and maneuvering in the environment, but with the increasing complexity of the abstract relationships that can be extracted by the system, and particularly with the development of symbolic representations, the process can become much more general and very powerful. This is a likely description of the fundamental mechanism underlying the experimentally well-documented "forward search" strategy that is at the heart of the human approach to problem solving [p. 205].

SOME BASIC MECHANISMS OF SCHIZOPHRENIA: WHAT ABNORMAL NEURAL CONTROL TEACHES US ABOUT NORMAL CONTROL

Schizophrenia is an "experiment in nature" that offers us an opportunity to understand the normal systems of neural control. Just as Kent begins with the concept of perceptual sets, the neuropsychiatric schizophrenia research group at Tokyo University (Utena, 1979, 1984; Saitoh et al., 1984a, b; Hiramatsu et al., 1986; Niwa, 1989) employs the same conceptual framework. N100 and P300 (P3) are evoked response potentials, that is, baseline brain wave changes that are studied because they seem to reflect the functioning of selective attention. The evoked response potentials and reaction times of schizophrenics and normals are compared. Abnormalities in late positive components of event-related potentials have been seen as proof of a genetic predisposition to schizophrenia (Saitoh et al., 1984a). Although the P3 evoked response potential latencies under certain circumstances are almost identical in both groups (Hiramatsu et al., 1986), generally schizophrenics and their probands show abnormal P3 (Saitoh et al., 1984b), and generally, the reaction times (RT) are also slower in the schizophrenic group. Most interesting is the observation of Fukuda et al. (1989) that under the correct conditions (mostly involving encouragement), abnormalities in P3 in schizophrenics can actually be improved. These abnormalities can be interpreted as (1) supporting the concept that there is a nonspecific deficiency in information processing in schizophrenia and (2) favoring the idea, originating with Donchin (see Donchin, Ritter, and McCallum, 1978) that this difficulty with P3 is the specific result of the experimental subjects having trouble updating or maintaining models of the environment in working memory. Readers interested in the details of this work at Tokyo University will wish to consult the references cited earlier. For our purposes, however, we will focus on the work of Niwa (1989), who has participated in the Hiramatsu, Saitoh, and Fukuda research but has written speculatively about possible neural mechanisms. Particularly relevant to this discussion are Niwa's views on the mechanisms of neural control involving the prefrontal cortex.

Niwa's paper[1] deals with mind– brain relationships. He explains as follows:

> By the expression "brain language" I mean the operational language or, in other words, the grammar or operation system of the brain for information processing . . . The disorder of schizophrenia is probably a disorder of this brain language system.
>
> We have worked on a model [see Figure 3] in which brain information processing systems recognize stimuli and react. What we call the "operation set" in this figure refers to something which loosely overlaps what we call the "imagination set." To elaborate, the "stimulus set" (meaning the stimulus processing set) is divided between the identification of particular characteristics of the stimulus information, a comparison and reference to what is stored within memory, and an effort at understanding meaning. "Response set" (the response processing set) is divided between the process of selecting a reaction, locating an appropriate association from memory, and the analysis of the output system independent of context and the individual processing function [p. 87].

Niwa elaborates the most difficult part of the model, which defines the "imagination set" and differentiates it from the other functions:

> "Operation set" (also referred to as the system control set) and what we are calling the "imagination set" (referring to "mind") can also change processing functions of the stimulus set and the response set . . . In other words system controlling sets [both within and without volitional control] can organize and supervise the proper action [decisions] within and recognition of the subject's environment. This allows for improved organization and reaction disposition along the lines of controlling the organism's entire activity, and furthermore allows for change through feedback of both [stimulus perception and response decision making] processes [p. 87].

[1]The author takes full responsibility for the translation of Niwa's research that appears here.

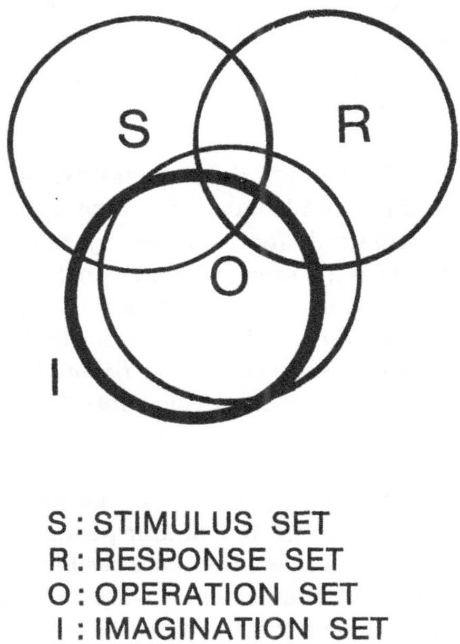

Figure 3. The model for the fundamental mechanism of personal cognition and behavior. (From Niwa, 1987).

Niwa is making a distinction between mental and brain-related functional systems, but an obvious ambiguity remains to be clarified. He has suggested that the imagination set acts by means of its affect on the stimulus and response set, but he has not yet fully explained his idea of how language and information processing are connected. He provides this explanation in the ensuing discussion.

> While on the one hand the operation set function is controlled by the brain's language and is both natural and autonomous, on the other hand the imagination set's function (the language of mind) is controlled by [the learned rules of] one's natural language (Japanese, English, etc.), i.e., by the manipulation of symbols of ordinary meaning, and through visual imagery stimulated by this process. In the case of the human brain, which is so highly developed, there is a need for the highest level of natural language control in order to provide for effective information processing . . . The functions of language are various, and include development of social relations, communication, etc. but the grammar of the language is [most]

> important since this is the means by which natural language controls the brain. *Through the use of this grammar, the mind's language is changed into a vocabulary of basic operating instructions for the brain*[2] [p. 87; italics added].

Niwa notes that although we do not understand the "machine language" of the brain, this brain language (as he puts it) and the mind's language are clearly connected, for example, by means of a shared grammar. That is, they share the same organizational rules. To him, this connection is of critical importance in understanding the puzzle of schizophrenia. The splitting in this disorder has been described by Hiramatsu et al. (1986) as, most probably, secondary to a decoupling between the stimulus process and the response process in central information analysis. The decoupling would explain the electroencephalic abnormalities in such patients, as well as their prolonged RTs. However, Niwa is trying to make a novel suggestion that goes beyond this earlier formulation. He continues his analysis by stating that one of four possibilities exists in the schizophrenic: (1) Schizophrenics demonstrate a response-set difficulty in which they fail to make use of information that becomes too complex to handle; (2) they have a problem with the stimulus set, which creates "a lack of functional stability in information analysis" (p. 88); (3) they fail because of an operation-set problem in that there is a decoupling of stimulus and response set systems; or (4) they have some nonspecific, undefined problem in taking action (which could be considered a variation of response set difficulty).

I believe, however, that he makes a fifth suggestion as well: that imagination can affect the operating system of the brain by imposing erroneous rules on the grammar of both the mind's and the brain's respective but related languages. I think this novel solution deserves to be noted and its implications considered. Moreover, Niwa goes on to explain that the essential problem seems to be within the core system for control of stimulus–response analysis. In his opinion, this means the failure of some neural control function within a division of the frontal cortex.

[2]What this internal grammar consists of, and how it works are exceedingly important questions for psychiatry in general. We know that individual experience influences brain development. But there is research by Tsunoda (1987) that suggests that the specific language that we use may affect the way our brain is lateralized for different brain functions. For example, although westerners habitually show right hemisphere predominance when listening to vowel sounds, the sounds of the human voice (laughter, crying, humming), the sounds of insects, etc., Japanese people (and those westerners fluent in the Japanese language) lateralize these sounds to the left hemisphere (p. 84)!

Believing that drugs that treat schizophrenia effectively also have an impact on the prefrontal cortical system, he hopes for the elaboration of this mechanism. In other words, *the frontal/prefrontal cortex is the master regulator of input/output relations within the brain.*

Lest anyone think that the Tokyo University Research Group's findings implicating the prefrontal/frontal cortex in schizophrenia represent a minority viewpoint, it is important to note that hypofrontality in this illness has been supported by work in BEAM studies, regional cerebral blood flow (rCBF) studies, and PET scanning as well (see Andreasen, 1989; Ingvar and Franzen; 1974; Mathew, Duncan, Weinman, and Barr 1982; Weinberger, Berman, and Zec, 1986; Berman Zec, and Weinberger, 1986; Morihisa and Weinberger, 1986; Buchsbaum, De Lisi, and Holcomb; 1985).

CEREBELLAR RESEARCH ON NEURAL CONTROL

According to Itoh (1988) "the exact mechanism of voluntary movement control is still unknown, but based upon . . . postulates of cerebellar-aided feedforward control, one may assume the following scheme for the implementation . . . " (p. 8). This example concerns arm trajectory formation, but I believe the implications go beyond control of the motor system and relate to our original question about general neural control. The sequence is as follows: the trajectory for the movement is first conceived within the cerebral association cortices. This trajectory plan is next fed to the premotor or motor cortex, or both, which "convert the planned trajectory into command signals" (Itoh, 1988, p. 8). The signals finally impact on the skeletomuscular system to produce the intended movement. During this process, there is both feedforward and feedback control as follows: (1) the peripheral effects, that is, the actual movement and its effects, provide feedback; (2) the cerebellum provides feedforward by means of cerebellar-corticonuclear microcomplex activity. (See Figure 4, describing the control "wiring" arrangements provided by this cerebellar mechanism.) Important here, however, is that this cerebellar control component is then capable of gradually taking over the function of the executive cortex. As Itoh puts it, on one hand, the cerebellar system internal feedback loop would be formed, which "would then replace the feedback loop through the external world" (p. 9). On the other hand, this "cerebellar dynamic model" (what I refer to in chapter 3 as Itoh's self-in-the-world model) is fed by the cerebral cortex command signals and sends its output back to the cerebral cortex.

What we are describing here are critical components of the machinery that the brain uses to perform the highest level of its functioning. These are the same complex functions that we described at the beginning of this chapter as properties of the prefrontal/frontal cortex. Clearly, the

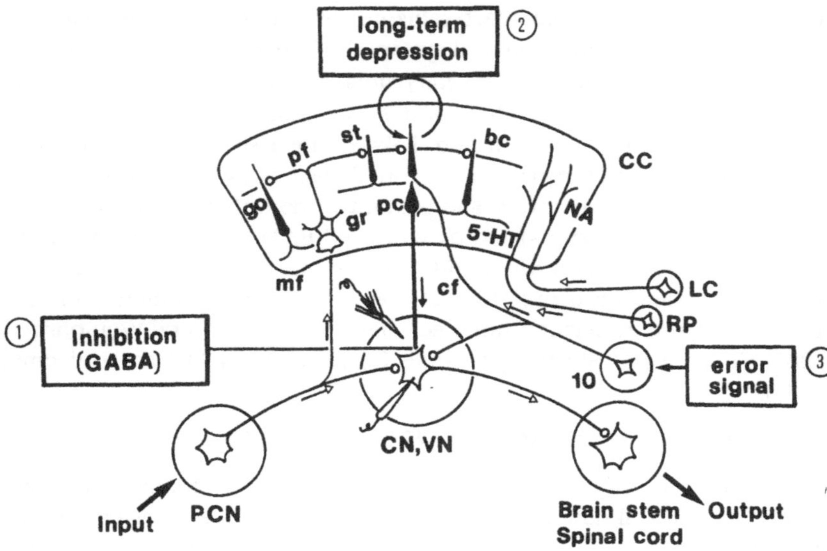

Figure 4. Structure of a cerebellar corticonuclear microcomplex. (From Itoh, 1988). CC = cerebellar cortial microzone; CN, VN = cerebellar and vestibular nuclei; PCN = precerebellar nuclei; IO = inferior olive; LC = locus ceruleus; RP = raphe nucleus; NA = noradrenaline; 5-HT = serotonin; mf = mossy fiber; cf = climbing fiber; pc = Purkinje cell; bc = basket cell; st = stellate cell; gr = granule cell; go = golgi cell; pf = parallel fiber; 1,2, and 3 = major findings that suggest adaptive operation of the corticonuclear microcomplex.

delineated neural control system for a self-in-the-world model includes a number of major parts: (1) the cerebellum, (2) the cerebral cortex (especially the prefrontal system), and (3) the basal ganglia. According to Itoh, learning will occur even when there is no actual movement involved, that is, imagined movement plans are also the expression of internal (metabolic) activity of the cerebellar-cortical system previously described, and will lead to learning just as though the imagined action had actually occurred! Support for this concept comes from Decety and Ingvar (1988), who show enhanced cerebellar activity (i.e., increased regional cerebral blood flow) during imagined tennis playing. Specifically, this learning pathway in which the cerebellum augments the cortical neural control involves the rubro-olivo-dentate triangle (see Itoh, 1988, p. 11).

But, as noted, a second system plays a different kind of role, in assisting neural control. According to Itoh (1988), this second system is the basal ganglia, massive structures that lie deep within the cerebral interior and that link the brain stem and cerebral cortex. Since lesions here produce such illnesses as Parkinsonism and akinesia, it is probable that the basal ganglia maintain stability of motor systems (p. 12). Thus, the cerebellum and basal ganglia contribute, respectively, augmented and stabilized neural control to the system that they make with the prefrontal cortex.

THE PREFRONTAL CORTEX AND SERIAL ASPECTS OF LANGUAGE AND SPEECH

Ingvar (1983 a, b; Ingvar and Franzer, 1974) and Lassen (1987) have helped establish some of the currently employed, noninvasive, real-time brain scanning technologies. Ingvar has exploited these new technologies for studying regional cerebral blood flow (rCBF) and, by implication, regional cerebral metabolic rate (rCMR) for a number of purposes. For example, the work of Decety and Ingvar (1988) tends to confirm the work of Itoh regarding the role of the cerebellar circuits in neural control.

In a selective review, Ingvar (1983b) notes that

> prefrontal dysfunction . . . in cerebrovascular disorders, organic dementia, Parkinson's disease, and schizophrenia are accompanied by various types of prefrontal rCBF and rCMR abnormalities. In such states different forms of "aseriality" of speech can be recognized which include non-fluent forms of aphasia of the Broca type, the hesitant, reduced and aprosodic speech in organic dementia and Parkinson's disease, as well as the peculiar semantic and motor disturbances of speech in schizophrenia [p. 2].

According to Ingvar, there is reason for believing that the prefrontal cortex handles the serial programs of the CNS, which are required for semantic and motor action programs for speech and language. Both the semantic and the motor parts of language and speech "emanate from serial action programs, that is, highly complex neuronal events that are firmly organized on a time basis" (p. 3). Ingvar feels that these are partly determined genetically (Chomsky, 1972, 1975) and partly acquired by experience.

What Ingvar is stressing by using the phrase "serial action program" rather than "conceptual structures" is both the seriality, "i.e., the temporal organization—at various levels of complexity—of the neuronal speech functions. . . " (p. 3), and the computational aspect of this function as well (p. 4). He feels there is an advantage in using rCBF and rCMR scanning technology (instead of EEG or ERP), since the neuronal generators of electrical events recorded by the scalp electrodes are currently unknown, whereas the scanning technology identifies the structures involved unequivocally (Ingvar, 1983a, b).

Regarding the prefrontal cortex itself, Ingvar, (1987) states that its role as a serial processor is a concept that has slowly emerged in the neuroscientific literature. Ingvar (1987) and Fuster (1980) define the prefrontal role in serial afferent and efferent functions; that is, it participates in the perception and programming of both *events* and their motor *responses* (each of which is temporally structured). Ingvar notes a number of major findings: (1) his observations fit with Luria's work, indicating that the collaboration of a number of different cortical functional units is required for the overall task of communication. Most important, (2) " . . . there is no clear evidence that specific circumscribed speech centers in the dominant (verbal) hemisphere are always activated when language and speech is perceived or produced" [i.e. Broca's and Wernicke's areas] (p. 6). (This unexpected result demonstrates how important it is to have a scanning technology by which one can test out hypotheses about brain functioning.) (3) Language and speech engage both hemispheres in a symmetrical fashion; and (4) the prefrontal cortex "plays a fundamental role for the interpretations of serial messages conveyed by language and speech" (p. 6). In fact, there is a possibility that "all forms of sensory perception and conscious awareness of the surrounding world takes place with the aid of serial motor and cognitive-action programs handled [i.e., decoded and encoded] by the prefrontal cortex" (Ingvar, 1983b, p. 26). Interestingly, Ingvar concludes that it is impossible to differentiate between pure language and purely cognitive functions (p. 27). A great number of detailed research measurements are presented by Ingvar to support his conclusions.

INTEGRATION OF SOME NEUROSCIENTIFIC AND ANALYTIC PERSPECTIVES

Gedo (1989) describes adaptation as "the attempt to fulfill as many as possible of [a] person's lasting motivations by means of selecting from a repertory of patterned modes of behavior those particular alternatives that have the best chance to attain the desired goals in the context of current circumstances" (p. 5). Clearly, a computational process is involved, and these "computations" have been a major subject of this chapter. Analysts necessarily consider the idiosyncratic factors that influence internal decision making (computing) in working with specific analysands. Such factors as affects, drives, conflicts, fantasies, and self-concept clearly constitute the most critical parts of the analyst's world, precisely because the analyst is charged with helping people with emotional problems. Cognitive psychologists and research neuroscientists, because of their focus on the general rather than on the particular, understandably think in terms of information processing (sans information processor).

Our task at the moment is to attempt a synthesis of these two fundamentally different approaches, particularly as they pertain to the subject of the prefrontal cortex and neural control.

We have surveyed four different research domains, each of which converges on the subject of neural control and the prefrontal cortex. From Kent (1981) we learned that this feedforward control system provides judgment and insight by carving out from the rest of the brain a system that selectively attends and that is capable of temporarily delaying, while using "snapshots" of previous experience, to make informed judgments about future action. From Itoh (1988) we learned that there is also prefrontal control over the basal ganglia, to stabilize the control system, and control over the cerebellum to augment the system for neural control. The cerebellum provides a self-in-the-world model, the manipulation of which can, at times, replace actions in the real world (this is what thinking as "experimental action" means). The various types of control, from an engineering perspective, have been diagrammed so that we can better understand them. From Niwa (1989) we learned that this prefrontal system serves as the basis for stimulus set, response set, and their coupling. Both Niwa's (1989) and Tsunoda's (1987) research also suggests that the "grammar" of *our mind's language bears a (control) relationship to the brain's language. That is, one's native language can alter brain organization (and certainly hemispheric lateralization).* Finally, from Ingvar (1987) we learned that brain scanning research confirms that the prefrontal cortex plays a role in the perceptual system, in the response system, and most certainly in the system for encoding and decoding speech and language. Ingvar particularly noted the role of the prefrontal cortex regarding "seriality".

The empirical findings of Ingvar support the speculations of Niwa (1989) not only about neural control, but also about the "language" of brain and of mind. If we consider Ingvar's research on the prefrontal cortex as a serial processor together with Niwa's work, it seems clear that the prefrontal cortex has a controlling influence over the internal "grammar," which is the deep structure by which neural control is managed. Niwa seems correct in asserting that the language of the brain and that of mind find linkage within the functional anatomy of the prefrontal cortical system for neural control. In other words, the logical problem-solving methodologies that Kent (1981) has described and the serial speech and language functions that Niwa (1989) and Ingvar (1987) have investigated seem to be merely different aspects of the same fundamental brain processes. Also, as we shall see, seriality (temporality) is a critical factor in synthesizing these research results.

Freud was, of course, well aware of the importance of the ability to delay motor output (action), so Kent's (1981) discovery of the use of "mental snapshots" is not novel. His descriptions are helpful, however, as an operational definition of what delay means in the case of adaptation. To Hartmann (1964) adaptive delay was the vaguer "partial replacement of merely reactive motor outlet, and of instinctual breakthrough, by directed and organized action [as] . . . an important part of ego development and an essential step in replacing the pleasure principle by the reality principle" (p. 39). Hartmann was the major explicator of this terrain of ego psychology, which describes how at some point in human development "a momentary pleasure, uncertain in its results, is given up, but only in order to gain along the new path an assured pleasure at a later time" (Freud, 1911, p. 223). If we now are able to go beyond Freud and Hartmann in this area, it is only because knowledge of the brain has grown enormously since the time of their research.

Hartmann raised the question of what constitutes rational or reasonable judgment in decision making. "What is called reasonable is actually frequently based in part on a set of implicit or explicit value judgments, the validity of which is taken for granted, and its meaning varies accordingly" (p. 46). He stated that

> in the history of philosophy many attempts have been made to link the concept of reason with definite mental functions. However, the rationalistic approach did not get very far in this direction . . . Also, the high evaluation of "reason" led many philosophers to believe in the actual or near omnipotence of intellect and to scotomize the true strength of the irrational factors [p. 46].

In other words, Hartmann made very clear that the relationship between rational and irrational behavior was actually too complex to understand (given the knowledge then available). His study of the decision-making process concludes with the idea that there is, in essence, a compromise in healthy personalities between decisions that involve the repetitive use of old solutions (which has a certain economy to it) and those that involve new solutions (with some apparent risk related to their novelty). Hartmann is trying "to describe *the relation of action to the setup,* [that is] the conditions existing in the psychic structure" [emphasis mine] (p. 63). Other factors involved in neural control, according to Hartmann, (1964, p. 60-62) are the use of "regression in the service of the ego," credited to Kris (1952), and the use of the "synthetic [i.e. organizing] function of the ego," as discussed by Nunberg (1930), which is seen as mediating between the three mental agencies of the tripartite model.

To return to the subject of seriality, or temporality, the brain's ability to remember the time relationship between events (i.e., their order within a hierarchy) is critical for a vast array of mental functions. For example, Boschan (1989) wrote that disturbances in this function are "a regular feature of the pathologies where narcissistic resistances [i.e., disturbances] prevail" (p. 261).

> These alterations in temporality may take different modalities [sic]: in the narcissistic pattern of NR [narcissistic resistance], what Anzieu (1975) calls "circular time" predominates; in the autistic type of NR, a fragmented, episodic time prevails, with important dissociations of temporal continuity; "each time is all anew," as a patient used to say. The countertransferential correlate of this type of distortion in the sense of time is the dismaying feeling that all that was analyzed in earlier sessions seems to have been completely wiped out, as if it never existed [p. 261].

Quite unpredictably for a psychoanalyst, Boschan adds:

> I thought it of interest that in neurophysiological concepts about attention, its twofold function is also noted: as a filter or selective barrier for stimuli, letting through the relevant ones, but also excluding those stimuli which are not relevant. In the recognition of relevance, the feedback loops have a central role (Callaway and Dembo, 1958; Hernandez Peon et al., 1961) to such an extent that an affectively meaningful stimulus immediately produces a reduction of the responses being evoked by a neutral one. Neurophysiologically it is considered that those stimuli reaching the mind through this selective barrier have a different mental inscription from those which arrive through other channels [p. 260].

What Hartmann considered to be adaptation, what he described in ego psychological terms (in a manner that was faithful to Freud's tripartite model), is the same problem current psychoanalysts, such as Gedo (1984a) and Boschan (1986) wrestle with in discussing modern intelligence theory and neuroscientific knowledge. But these are also the same issues that Kent, Itoh, Niwa, and Ingvar are studying, insofar as their research relates to how the brain makes complex decisions requiring insight and judgment. Only the terminology is different. Furthermore, as noted by Boschan (1989) as well, it is clear that (1) the adaptive decision-making process indeed involves selective attention, which functions as a filter or selective barrier, but also (2) temporality (or seriality) seems to serve as the fulcrum for a vast array of neural functions required in adaptive strategizing. Both functions, selective attention and seriality, are provided by the prefrontal cortex.

The role of seriality in mental processing deserves a separate chapter, but for our purposes it is sufficient to note that the arrangement of goals and action plans into hierarchies (based on the retention of correct time order of memories of the analysis of goals), and the use of language (the encoding and decoding of which is also time dependent) depend on an intact prefrontal cortex. We can begin to see why damage to this part of the brain produces the broad-ranging and profound disturbances noted at the beginning of this chapter. Because an important part of the self is connected with language or with hierarchies of goals and values, without the ability to rank order internal memories according to the time dimension or to communicate, the very experience of self would obviously become profoundly disturbed.

Let us summarize more broadly the significance of prefrontal cortical neural control to psychoanalysis:

(1) At the level of clinical theory, it should become possible to understand better a number of specific illnesses, including such severe ego disturbances as schizophrenia, because input, output, and coordination of goal direction are likely affected by prefrontal injury in this specific illness. It should also be possible to create a better nomenclature of schizophreniform illnesses, on the basis of our knowledge of which subsystem of the prefrontal cortex (or other areas) is dysfunctional in specific cases. Psychoanalysts seem to be in a preferred position, for the future, by virtue of their relationship with individual patients, to help neuroscientists test out hypotheses such as the aforementioned ones. Patients with schizophrenia or other disorders could be studied simultaneously by neuroscientists using rCBF and rCMR methods and by psychoanalysts using psychodynamic treatment methods. Thus, the two quite different approaches might integrate and cross-fertilize each other.

(2) At the level of diagnosis and treatment, if we appreciate that the prefrontal cortex accomplishes its neural control through a number of specific computations, we are in a better position to facilitate our patients' progress by observing the exact nature of their difficulties. For example, we might learn to differentiate clinically between trouble with the encoding or decoding of perceptual sets, trouble identifying reasonable goal states, and problems in selecting appropriate responses from a hierachy of plans. More specifically, we may be able to detect a category of patients (predictable from Ingvar's research) who show an inability to experience or remember the order and, therefore, the "logic" of their experience. If such people were to seek psychiatric help now, the probability is high that this neurocognitive wrinkle in their psyche might not be appreciated.

(3) Another kind of patient with a different profile of prefrontal dysfunction, is worth mentioning. These are patients whose judgment is impaired specifically because they can not use "mental snapshots" while temporarily postponing a response; (they have a long-term memory re-

trieval disturbance. In fact, Koh and Kayton (1974) showed that schizophrenics in particular suffer from such a retrieval problem. As the reader will recall from Kent's (1981) description, this step of "looking before leaping" is decisive in preventing catastrophic events; it is the moment when we develop a mental picture of what might happen, when we tap long-term memory stores to see if what is perceived connects with, or matches, any earlier experience of danger. Some people are incapable of executing such planning delays, and although their behavior looks masochistic in the sense that it is self-defeating, it is not really intended to cause misery on some deep unconscious basis. Rather, their behavior might simply be called "failed behavior" in that it is imperfectly conceived. I noted this possibility during a panel on sadism and mascochism in neurosis (Levin, 1989b), and Kernberg was highly in agreement that all self-defeating behaviors are not "masochistic" in the psychoanalytic sense.

I could go on enumerating various specific prefrontal dysfunctions and how they might manifest themselves in a psychotherapy or psychoanalysis. However, it is sufficient just to have made the point that prefrontal dysfunction can occur and that it needs to be identified as such. Having worked with several neurocognitively damaged persons, I know how difficult this task can be.

(4) At the level of metapsychology, it is important that we take into account the knowledge we have acquired about the brain's system for neural control when we are building our psychoanalytic theories. Gedo (1989) did so in a revision of his hierarchical model. How we understand such variables as judgment and insight depends on how we frame these phenomena in our minds, whether in terms of Freud's tripartite model or with awareness as well of the tripartite system of prefrontal cortex, basal ganglia, and cerebellum, and their participation in neural control. Or possibly we will keep both frameworks in mind and find evidence clinically for oedipal conflicts or neurocognitive disturbances when they are relevant. The last word has not been written on the optimal psychological units from which behavior can best be understood. Our psychoanalytic conclusions will be on the surest footing only if we first take the time to observe and form detailed judgments about how our patients process information, and for this we need to appreciate what is being accomplished in cognate disciplines.

SUMMARY AND CONCLUSIONS

This chapter has presented some complex ideas integrating the neuroscience concept of neural control and the psychoanalytic concept of adaptation. This integration allowed a kind of microscopic analysis of adaptive mechanisms. Neural control seems to be under the regulation of the prefrontal cortex, which recruits other brain areas, such as basal ganglia and cerebellum, into a system that provides insight, judgment, and selective attention. These terms have been operationally defined within the context of the prefrontal brain system.

Breifly stated, decision making involves a definable sequence: the prefrontal cortex influences selective attention to an aspect of current experience relevant to the decision that needs to be made, by means of its influence on such perceptual-motor structures as the frontal eyefields, which tell our eyes where and how to look; it delays cortical motor discharge temporarily, by way of its effect on the pons; during this delay period, information in the form of mental snapshots of current perceptual events are then compared with similar experience stored in long-term memory to find signs of danger; only then is a decision made. The prefrontal cortex brain-operating system also seems to be organized to some extent around the sense of time. This function of seriality, or temporality, not only is part of the prefrontal cortex's ability to encode and decode language and speech, but also seems part of the brain's "machine" language as well (the language it uses to communicate with itself).

The converging evidence from several disciplines reviewed in this chapter supports the concept that the language of mind and of brain are interconnected by the grammar rules that they appear to share, for example, within the prefrontal cortex. Niwa (1989) has suggested that our natural language provides an entré into the brain's (operating system) language. And Tsunoda (1987) has shown that our language can alter such an important variable as the lateralization of the brain for sound perception. The "grammar" in our brains is thus partly genetic and partly the product of experience. It is the latter that allows for psychoanalytic intervention. We have discussed some ways in which adaptive decision making might fail and how this might present itself clinically. Finally, we have considered the impact such neuroscience perspectives might have, or are already having, on psychoanalytic metatheory (Gedo, 1989).

5

The Hierarchical Developmental Model

Neural Control, Natural Language, and the Recurrent Organization of the Brain

PRÉCIS

Chapter 5 formally introduces the reader to the developmental hierarchical model of Gedo and Goldberg (as recently modified by Gedo), as well as to Gedo's theorizing on the subject of development. I discuss some details and implications of Gedo and Goldberg's model throughout this book. This chapter focuses primarily on the structural similarities between the hierarchical model and the prefrontal cortex in unifying and goal-directing activities. Picking up on insights expressed in Chapter 4 on possible relationships between the language of "brain" and of "mind," the present chapter emphasizes the organizing function of formal, native language exposure.

Because of the complexity of such relationships, Chapters 7 and 11 also deal with aspects of language, as do many other sections of this book. As far as language functions are concerned, the brain is organized at various levels: (1) the level of formal or native language; (2) the level of cognition (difficult or impossible to distinguish formally from language activity per se), which might be considered to provide a "software" or programming (language) function; and (3) at a very basic operating system level the hardware or "machine language" of the brain. None of these levels of activity is thoroughly understood. What is exciting to me is how the work of Gedo, a neuroscientifically informed psychoanalyst, and the work of many neuroscientists (often informed, but relatively less interested in psychological issues) dovetail with each other.

Equally meaningful to me has been the opportunity to teach and lecture in two cultures (in the United States and in Japan), which has made it easier for me to move between and attempt an integration of the two major scientific weltanschauungen *of this book: neuroscience and psychoanalysis. In attempting to translate from one field to another, or one language to another, one is forced to explore more thoroughly how specific observations and theory are related to each other. Theory formation and observation are each much more highly culture bound (and therefore arbitrary) than we ordinarily recognize, and this often significantly limits our perceptiveness. As is noted in chapter 8, Einstein was therefore mistrustful at times of both deductive and inductive methods.*

In many scholarly papers, first with Goldberg (Gedo and Goldberg, 1973) and then alone, Gedo (1979, 1981, 1984a, 1986, 1989a) has elaborated a model of psychological development that is both elegant in its simplicity and yet in keeping with the veritable explosion of knowledge within the neurosciences. A mere handful of psychoanalysts (including Reiser, 1985; Schwartz, 1987; and Lichtenberg, 1988) have engaged the difficult task of balancing the insights between fields rather than deciding in favor of paradigms of a single science. In what follows I describe what I consider the core of Gedo's hierarchical model. I wish to expand as well on the relationship between this model and two fundamental phenomena; neural control and the languages of mind and brain. The reader may then appreciate more deeply how the Gedo model cuts across disciplines by identifying the importance of hierarchies in the brain and mind and the criticality of language in general and psychoanalytic interpretations in particular in decisively altering man's inner (hierarchical) organization.

It may be helpful,if I briefly locate Gedo's theorizing within psychoanalysis. One might arbitrarily divide post-Freudian psychoanalysis into four major theoretical camps: drive-defense, self-psychological, hermeneutic-philosophical (linguistic), and a systems or interdisciplinary (eclectic) approach. Drive–defense theory focuses primarily on conflict and its resolution and is based on work with neurotic people. This school is most often assumed to be representative of the field of psychoanalysis (at least so far as analysts describe their work in writing). Anna Freud and Heinz Hartmann were the two pioneers most connected with this school, and the evidence is that they saw themselves more as preserving, rather than revitalizing or modifying, Freud's original theorizing. Self psychology can be traced from Freud's (1914) "On Narcissism," through the dissidence of Jung and Adler, to Freud's later interest in the application of psychoanalysis beyond the diagnosis of neurosis: first to children and then to a heterogeneous cluster that has included psychotics, those who would be called borderline or narcissistic personalities today, and such other kinds of characterologically disturbed individuals as addicts and perverts. Self psychology draws on the work of Melanie Klein, Anna Freud, Fairbairn, Winnicott, and the British object relations school, as well as on the contributions of such Americans as Harry Stack Sullivan, Otto Will, and Ping Nie Pao. The hermeneutic (or philosophical) school propounds sensitivity to the meaning of the patient's personal sense of history and communicative style. Freud was interested in language as a deep structure of the brain (e.g, consider his "On Aphasia"), and his lead has been explored by Saussure, Jakobson, Rosen, Derrida, Lacan, Recouer, Ornston, Mahony, and others.

I have saved the systems (i.e, neuroscientific or eclectic) school for last because it is to this branch of psychoanalysis that I assign Gedo. Of course, the division of psychoanalysis just outlined is arbitrary to some significant degree; for example, Gedo is clearly also interested in language. The different schools seem merely to express their personal preferences among the multiple factors that Freud felt contribute to mental life: sexuality, aggression, and conflict; disturbed narcissism; language and personal history; and neuroscience. It is probably fair to say that the formal descriptions of the tenets of the various schools are more likely to differ from each other than the actual practice of such views. Thus, seasoned and talented clinicians will at different times make use of the insights of each of the perspectives, depending on what is appropriate or expedient. How even subtle theoretical differences might affect therapeutic outcome remains to be studied systematically.

Of what does the eclectic or systems approach consist? Gedo presents what he considers to be the core of a credible, internally consistent psychoanalytic theory supported by three distinctly different sorts of evidence: clinical psychoanalysis; infant and child observation studies; and contemporary neuroscience. At the core of the neuroscience contribution, as I interpret it, is the work of Jackson of a century ago, demonstrating clinically the fundamental hierarchical organization of the brain.

Dispensing with libido theory, ego psychology, dual-instinct theory, and object relations theory as such, Gedo builds on the following sources: (1) Anna Freud's conceptualization of multiple lines of development (an extension of her father's simpler concept of the developmental sequence of autoerotism to narcissism to object love), which represents a unique organization of psychoanalytic data in the direction of a systems approach and is fundamentally included in Gedo and Goldberg's (1973, p. 7) hierarchical model. Ludwig von Bertalanffy and Roy Grinker, Sr. must also be mentioned for the former's development of his general systems theory and for their joint effort in employing this framework to psychiatry; credit goes as well to Erikson for pioneering epigenetic models; (2) Piaget's epistemology of knowing described by Rapaport as "a hierarchical series of thought organizations [which] arises, in the course of maturation and development, culminating in reality-adequate thinking" (quoted in Gedo, 1986, p. 65); (3) Melanie Klein's recognition of the role in neurosogenesis of the failure to integrate ego/self nuclei, which is also adumbrated in Gedo's model (see the discussion of Klein in Gedo, 1986, p.82–98); (4) Ferenczi's description—which according to Gedo (1986, pp. 40–41) was years ahead of his time—of narcissistically injured or borderline individuals, their fluctuating ego states, and the role in their lives of failures in good enough mothering. (Gedo has noted that subsequent research in self psychology—including Kohut's, most prominently—has moved, without derogating oedipal-level pathological mechanisms, toward the kind of integration of interactional and intrapsychic approaches contained in the hierarchical model); and (5) the work of Basch, which has been decisive for Gedo, especially for its integration of vast numbers of infant observations and neuroscientific and philosophical knowledge into the corpus of psychoanalysis (other contributions, for the interested reader, are reviewed in greater detail in Levin, 1989).

THE HIERARCHICAL MODEL: ITS CORE

Gedo's (1986) current model, in contrast to the original version, developed with Goldberg (Gedo and Goldberg, 1973), strips the hierarchical model of its formal linkage to the five older models of which it was an amalgam. What remains are multiple, linked nodal points of development in which the organizational axis changes as a function of time. In this hybrid model one passes through the following stages: a stage of ego nuclei needing unification (i.e., a stage of nuclear or cohesive-self formation delineated as self definition, or Mode I); the establishment of self-awareness with the ability to formulate wishes and establish priorities among them (self-organization, or Mode II); a stage characterized by the ability to keep in mind differing (potentially competing) wishes and to deal adaptively with such conflicts and their associated affects (what Gedo calls self-regulation, or Mode III); the establishment of the ability to renounce wishes that threaten adaptive equilibrium (Mode IV, which coincides most closely with oedipal-level problems); and the achievement of a significant level of symbolic capacities (Mode V); and so on. (Note here that "and so on" indicates the possibility of adding nodal points to the hierarchy; i.e., the model is an open system.)

What is the essence of such a hierarchical model? Clearly, this will vary with one's theoretical perspective. As for me, the model's essence (aside from the arrangement of elements into a hierarchy that allows for progressive or regressive movements) is its conception of *the changing fulcrum or motive around which development is seen to organize at each stage.* These motives include the need to avoid overstimulation; the need to organize around real object relationships (with the focus on awareness of self and other and attunement concerns); concern with critical self-object functions; id/ego/superego concerns; and focus on adjustment of one's hierarchy of goals and values. Levey (1984/1985) has cogently described this developmental view and its implications for psychoanalytic metatheory in a paper on the concept of structure in psychoanalysis.

The implicit reason for the changes in underlying motive or structure in such a model of development is that the genetic blueprint for human beings is so written. Within genetics there has been the discovery of the "homeobox"(see DeRobertis, Oliver, and Wright 1990), that part of the genetic material that is conserved over a large number of phyla and that is the DNA master control for the general order in which development (gene activation) proceeds. What differs between phyla with respect to this general order of gene activation involves the nature of the structure that develops; for example, in an insect the plan specifies that wings are added to the step following body formation whereas in humans , limbs are the next addition. Of course, it remains to be learned how environmental experience may alter the expression of such a genetic blueprint. That such sequences are influenced by both biological (i.e., built-in) plans and experience (accidental or intended factors) is a crucial point that will be considered toward the end of this chapter. (This was also discussed in chapter 3, from the perspective of brain plasticity.)

THE HIERARCHICAL MODEL AND NEURAL CONTROL

Research on the brain is changing so rapidly that it is becoming extremely difficult to write about the subject with any certainty that what is said today will hold for tomorrow. In this sense, I value the opportunity to elaborate on some aspects of adaptation that I have written about elsewhere (within the framework of Gedo's model), since new information is now available that helps us better understand the relationship between adaptation as a psychological construct and adaptive change in a neuroscientific sense.

Adaptation and adaptive change are complex subjects. Gedo (1989a) has described adaptation as "the attempt to fulfill as many as possible of [a] person's lasting motivations by means of selecting from a repertory of patterned modes of behavior those particular alternatives that have the best chance to attain the desired goals in the context of current circumstances" (p. 5). Yet the question remains: How do we accomplish such a task? More specifically, how can we combine what we intuitively and introspectively know about adaptation with what is known about how the brain works so as to understand the fundamental mechanisms of such a psychological function?

Elsewhere (chapters 2, 3, 5, and 6, this volume) I describe ways in which the brain might facilitate adaptive decisions. I focus primarily on two basic mechanisms: (1) an adaptive integration of the two cerebral hemispheres in which a strategy is employed such that (a) interhemispheric communication blocks can at times serve as psychological defenses and (b) the matching of the brain subsystem best suited for the task with the problem at hand involves critical decision making; and (2) the provision by the cerebellum (more exactly, the vestibulocerebellar system) of decisive input to the cortex in carrying out many tasks, allowing adaptation to occur by means of either real or imagined actions (i.e., through what might be called "*gedanken* experiments" by way of manipulation of a cerebellar-based "self-in-the world model)[1]. Left unanswered in this earlier work, however, was how (i.e., where in the brain) such complex internal decisions are regulated.

Significant progress has, however, occurred (see chapter 4). I shall briefly summarize some of the conclusions here, since they bear on our understanding of Gedo's hierarchical model. What follows is divided into three sections: The first describes some of the evidence bearing on the connection between neural control and the prefrontal cortex. The second considers evidence relating neural control to the brain's temporal (serial) organization. The third speculates about the hierarchical organization of the brain's "operating system" and the role of natural (native) language in modifying the rules of this system.

First, however, a word about the hierarchical model and neural control. Gedo's model assumes that various mental functions and the memories of the situations out of which they arise are organized in hierarchical form. This hierarchical organization is what allows progres-

[1]The reader may better appreciate the significance of such models by reading Bower and Morrow (1990), who state: "We build mental models that represent significant aspects of our physical and social world, and we manipulate elements of those models when we think , plan, and try to explain events of that world. The ability to construct and manipulate valid models of reality provides humans with our distinctive adaptive advantage; it must be considered one of the crowning achievements of the human intellect" (p. 44; also see Boden, 1988). From the perspective of psychoanalytic theory, the cerebellar self-in-the-world model is the equivalent of what Gedo has called self-organization. In Gedo's Mode I, this model is rudimentary; that is, the nuclei of self are inadequately coordinated. Coordination of percepts improves in Gedo's Mode II. Finally, in his Mode III "characteristic motivations and/or affective patterns are also included in a coordinated gestalt" (Gedo, 1990, personal communication).

sive and regressive swings along the developmental axis of the model (as Jackson, 1958, understood such change). In other words, new developments usually, but not invariably, become superordinate; they come to represent preferred values, goals, or methods for achieving goals. I say not invariably because Gedo has stated that some development occurs outside the usual order; for instance, creativity requires this kind of "splitting" (see Gedo, 1989a). This aspect of Gedo's model appears to be the least detailed though it could turn out to be crucial for the model. There is an ambiguity regarding the rules governing the shift between modes within the model. One problem that suggests itself is Gedo's emphasis on individual control over the arrangement of elements within the various hierarchies, (and also control over the time sequence for their unfolding). But nowhere in Gedo's writings is there an explanation for the mechanisms of such control. (One possible mechanism, discussed later in this chapter as well as in Levin, 1989, is that the brain's operating system instructions for the unfolding developmental sequence are capable of being readjusted decisively on the basis of input from one's natural language). Of course, the possibility of individual control over the developmental trajectory does not mean that genetics has now been overruled. It only means that we need to think of the ways in which the genetic machinery is activated (or inactivated).

Some clinical material may clarify what is meant by the concept of language decisively changing neural organization. Let us first consider briefly the situation of those who are born deaf. Some deaf persons grow up to be intellectually and emotionally advanced adults, fully on a par with the healthiest hearing people. But others suffer from a critical linguistic-cognitive deficiency that significantly limits their ability both to conceptualize and to communicate, even by sign language. The advanced group of deaf persons can abstract, that is, "shift to categorical, definition-based lexical organization" (Sacks, 1989, p. 108); the latter or "slow" group are limited to perceptually-based organization and will show what experts in the field call "low language ability" (i.e., they may become functionally retarded). Sacks has cogently described these differences and has offered an explanation. His suggestion is that deaf children who are not exposed early to good language or communication may suffer "a delay (even an arrest) of cerebral maturation, with a continuing predominance of right hemisphere processes and a lag in hemispheric 'shift' " (p. 110). In other words, without exposure to signing or meaningful language input before puberty, these children will not experience the normal shift to left-hemisphere dominance, with its superior syntactical and abstracting abilities (Collins, 1990). Clearly, language acquisition can decisively alter neural organization.

This experience of some deaf persons who have suffered because of a lack of proper and timely language input is a very real and poignant problem of which I am well aware since I have been a consultant to a psychiatric clinic for the deaf since 1974. For this discussion, however, what appears crucial is that the experience of language deprivation in the deaf population supports the view that the establishment of one's natural (formal) language early in life is a developmental step that can be decisively organizing for the brain.

Although the emphasis here is on the possible releasing role of language exposure, development of advanced or abstract cognitive abilities is more complex by far than mere language input. Shallice (1988) describes a patient (Mr. R.), an accountant who had a large orbitofrontal meningioma removed. Six years after surgery his IQ was over 130, and he did well on a variety of psychological tests. However, his ability "to organize his life was disastrously impaired" (p. 336). He drifted through a series of jobs, fleeing from each; unable to manage his financial affairs or his marriage, he suffered bankruptcy and divorce. He could not be punctual or organized, nor listen to advice. Decision making became extraordinarily difficult, and he would take hours reviewing the details of such relatively trivial issues as the relative merits of various restaurants, so that he could decide where to eat. Shallice sees Mr. R. as lacking a supervisory organizational function. Without this capability, Mr. R. is trapped in a sea of irrelevant details within which he endlessly perseverates. "The primary function of the supervisory system is that of producing a response to novelty that is planned rather than one that is routine or impulsive" (P. 345). Thus, there is evidence that under normal circumstances the frontal cortex provides decisive organization by way of selective attention and selectivity of response. For this reason it will help to review research on the frontal part of the brain.

NEURAL CONTROL AND THE PREFRONTAL CORTEX

Several converging lines of evidence suggest that the prefrontal cortex regulates the brain's system for neural control, or what might be called (analogous to computers) the brain's operating system. This subject has been reviewed elsewhere (Benson and Stuss, 1989; Levin, 1989) but will be summarized here briefly. One line of evidence deals with the experience with persons who have undergone prefrontal cortical injury. This group suffers from a spectrum of difficulties, which may include concreteness, impaired attention, difficulty abstracting, problems excluding what is irrelevant, diminished spontaneity of speech, perseverative behavior, apathy, indifference, shallowness, and a kind of inappropriate jocularity called *witzelsucht* (Andreasen, 1989). The prefrontal cortex is also required for the coding and encoding of speech and language (Ingvar,

1987). Experiments with animals confirm that with injury to the prefrontal cortex there is a general difficulty in forming problem-solving strategies, especially using "selective attention" (Kent, 1981, p. 208). Memory per se is not the problem but, rather, there is an apparent disturbance in the ability of the animal to act on the memory of an object in the absence of its appearance (Kent, 1981; Godman-Rakic et al., 1983; Reiser, 1985). Most critical perhaps is poor "recency memory," that is, the memory for the time order of experience (Kolb and Winshaw, 1980). Ingvar (1987) reports the same time sense or seriality as a critical function provided by the frontal cortex, without which the language or speech function deteriorates.

A second line of evidence suggesting that the prefrontal cortex regulates the brain's system for neural control comes from studies of schizophrenia, which has been associated with several brain abnormalities on scanning. A group of right-handed schizophrenics showed diminution of the left temporal pole of the brain, as seen on computed tomography (CT) scans (see Levin, 1988, and chapter 10, this volume, for a review of this work of T. J. Crow). A second finding that has proven reliable in schizophrenics is prefrontal hypoactivity, as seen on magnetic resonance imaging (MRI) and in studies of regional cerebral blood flow (rCBF) and of regional cerebral metabolic rate (rCMR) (see Levin, 1989, for detailed references to the work of Andreasen, Ingvar, and others). The conclusion to be drawn from these studies is that the cognitive and language disturbances of schizophrenia are most likely the result of diminished prefrontal cortical functioning.

Studies of schizophrenia have also demonstrated alterations in reaction time (RT) and evoked response potential (ERP), the latter finding being reversible by an encouragement paradigm. Such work, in the United States and in Japan, has led to the speculation that the key pathogenic factor in schizophrenia, as indicated by the massive input–output processing (stimulus set–response set) problems, is likely the loss of prefrontal cortical control over the machine language (operating system) of the brain (Niwa, 1989).

A third line of research relates to the work on the cerebellum by Itoh, which I have reported on in chapter 3. Itoh's research and that of Kent (1981) and of Ingvar (Ingvar, 1987; Decety and Ingvar, 1988) suggest that the prefrontal cortical system for neural control is primarily a feedforward system with adaptive mechanisms (see Figure 1 in Levin, 1989) in which the crucial element coordinating the exercise of judgment and insight in decision making is the prefrontal cortex. The other elements of the system are the cerebellum and the basal ganglia, and the prefrontal cortex maintains this control by "sculpting out" (activating or recruiting) those other parts of the brain whose arousal is critical to the decision-making process. Note well that the prefrontal cortex also hastens decision making by controlling selective attention (by means of its influence over the frontal eyefields; i.e., we gaze on what we need to facilitate the decision). Research on the cerebellum is complemented by work in the field of artificial intelligence, which seeks to investigate by still different means the mechanisms of neural control (see Kent, 1981; Levin, 1989).

NEURAL CONTROL AND SERIALITY (TEMPORALITY)

As noted earlier, Ingvar has reported that the prefrontal cortex is necessary for speech and language encoding and decoding. Although language is usually thought to be mediated by Broca's and Wernicke's areas in the left hemisphere, Ingvar (1987) reports that studies of regional cerebral metabolic rate (rCMR) and regional cerebral blood flow (rCBF) demonstrate unequivocally that these areas are not activated in all language activity, whereas the prefrontal cortex is! What is of special interest to this discussion, however, is the role of seriality (temporality), since the arrangement of elements into a hierarchy within the brain would appear to require the ability to organize memories according to time or some other similar function. Now, it might be that the brain does not employ a time tag as a basis for distinguishing recent from remote memory, but this seems unlikely. That is, it seems probable that the brain's ability to distinguish memories in a time sequence (recency) is required for the hierarchical organization of virtually anything. Otherwise, recent solutions and old solutions would be indistinguishable from each other and would be randomly employed. Whatever we decide about hierarchies, therefore, should take into account the probability that this function requires an intact prefrontal cortex.

THE BRAIN'S OPERATING SYSTEM AND THE ROLE OF NATURAL LANGUAGE.

Mention has been made of the work of Niwa (1989), who suggests that the language of the brain and that of mind (i.e., our natural language) share a common grammar and that the latter influences the former. Put differently, our natural language provides us with an adaptive means of modifying the operating instructions of the brain (i.e., the operating system itself). We have also mentioned Sack's (1989) account of the world of the deaf, where language stimulation and acquisition seem decisive in modifying brain organization. But to appreciate fully the complex role of language, we must also mention the work of Tsunoda (1987), who studied lateralization of sounds using dichotic listening tasks. Tsunoda has shown that Japanese people, unlike Westerners, hear vowel sounds, human affective sounds (laughter, crying, humming), and the sounds of nature (crickets, for example) with the left, not the right, hemisphere. The most interesting of his findings, however, is that westerners who are fluent in the Japanese language also lateralize these listening activities to the left hemisphere! (Tsunoda has also studied Polynesians and Koreans, but these groups, unlike Japanese speakers, lateralize sound in the same manner that westerners do.) Tsunoda's research seems to provide us with the first convincing evidence of Niwa's speculation that one's natural language is capable of serving as a vehicle for altering the brain's operating instructions (i.e., altering brain organization for a specific task).

Tsunoda's research is important, suggesting as it does something about the language-based determinants of the brain's hierarchical rules of operation. But it is also vital to keep in mind that his research would be much less interesting if it were not for the findings of Niwa, Ingvar, Kent, Itoh, and their collaborators, who have helped generate important pieces of the brain operating system puzzle. Further, we owe a debt to Gedo and those who inspired his work, because these largely psychoanalytic efforts begin to place the basic research cited here in a meaningful psychological context, namely, that of human adaptation as a hierarchy of self-in-the-world potentials.

At this point it is worth speculating further on the nature of the adaptive process.

SOME ADDITIONAL SPECULATIONS

We cannot be sure how much of what humans are able to accomplish is actually "hard wired" into the brain and how much is the product of learning. Bharucha, studying musical discrimination, has demonstrated that much of what was considered innate musical "grammar" is really the consequence of repeated experience within a specific culture (cited in Heinrichs and Endicott, 1988). This finding should caution those who would make premature theoretical leaps about brain mechanisms. However, there does seem to be enough reliable empirical evidence to conclude that learning itself involves multiple systems and multiple mechanisms. At the system level, there are the corticolimbic, corticostriatal, and corticovestibulocerebellar systems, which complement each other and provide for the processing of the complex, cognitive discriminative kind of "declarative" learning, habit-pattern learning, and self-related episodic memory ("procedural") learning, respectively (see chapters 3 and 10, this volume). At the subsystems level, individual neurons are best understood in terms of the chemical cascade involving neurotransmitter(s) changes, 3′, 5′ -AMP (cyclic AMP), calcium channels, and activation of the operator gene. And although we do not yet understand the basis of long-term memory, there seems to be a consensus that it relates to changes in nucleic acid or related DNA compounds. The changes within each of these hierarchically organized, interconnected levels associated with experience constitute a comprehensive operational definition of learning. That is, learning is a characteristic of large populations of connected neurons, such as the hundred billion or so neurons (and their trillions of synapses) that constitute our brain.

But what more might be said about the relationship between natural language, learning, and the operating system of the brain (neural control)? A question comes to mind. *Is it possible that what we call natural language might be a design component within the brain with fractallike quality?*[2] *This would mean that the different modules (or levels) of knowledge within the brain*[3] *are interdigitated with each other by means of a recurring hierarchical arrangement of instructions that might be a shared property of both one's native (natural) language and also of the operating system of the brain.* Put differently, if, as with the "homeobox," there is a set of operating instructions for each knowledge system of the brain that specifies the "default" settings that obtain under ordinary circumstances, then there must also be a way of altering these default settings when circumstances warrant adaptive change. Perhaps language evolved to fulfill this task.

It may be helpful to describe the hierarchical nature of each level of brain organization. At the level of organization usually designated physicochemical, that of the DNA (mostly) in the nucleus of neurons, a clear hierarchical organization is demonstrated by structures like the "homeobox," which determines the order in which the various parts of the DNA blueprint are activated during development. At the next higher level of organization, which is usually called physiological, the brain is clearly arranged hierarchically in the form of multiple feedforward and feed-

---

[2]From the point of view of Mandelbrot's concept of fractal geometry, certain larger patterns in nature can be generated from the reproduction of smaller and smaller units of the same original shape. As an example of a fractal, consider a snowflake. Its six-sided symmetry has been shown to exist within a high power magnification of its component parts. Moreover, still higher power magnification of these parts shows that these microscopic snowflake components are made up of still smaller elements with the same six-sided snowflake symmetry. This recurrent quality is what defines fractals.

[3]Nadel and Wexler (1984) describe in detail what is meant by "modules," or knowledge-acquisition systems of the brain. The shift from associationist perceptions and narrow anatomical localizations for memory to the concept of subsystems specialized for specific tasks (facial recognition, spatial perception, language, etc.) makes a great deal of sense to me, and a paper is in progress on this important perspective.

back loops (see chapters 2 and 3, this volume; Levin, 1989). And finally, at the highest level of abstraction, that of neural control, at which level phenomena are usually labeled psychological, hierarchical systems in the form of the formal "languages" of the mind (native languages) and of the brain (its so-called machine language) once again serve as fundamental regulatory units. Therefore, *hierarchies are one of the key recurrent patterns inscribed into the brain.* From a neuroscientific perspective, Shallice (1988, following Luria) notes that "the triggering of schemata is frequently mediated by language in humans" (p. 333). Conventionally in psychoanalysis one's formal language is the level at which interventions are thought to occur that potentially alter the entire interlocking organization of the brain, as reported in the research paradigm of Tsunoda (1987) noted earlier. Finally, I have speculated here that one's natural language, once assimilated, permanently and decisively alters brain organization. Language may not only facilitate the development of the genetic plan for psychological organization, but it may also allow for adaptive reorganization as a solution to problems requiring novelty and for the manipulation of modules of knowledge.[4]

I have suggested (Levin, 1980) that metaphors in transference interpretations appear to contain coded elements that appeal to each of the three primary sensory modalities and that these cortical areas might become activated simultaneously (rather than serially), thus unlocking critical memories and opening the way for insights. Over the past decade I have continued to focus in particular on those areas of the CNS that are involved with cross-modal synthesis, for I believe intuitively that these sensory integration zones are likely to be the leading part of the brain's system for coding experience in abstract format. As suggested by the studies of Ingvar, Itoh, Niwa, and others noted earlier, it is possible that our natural language also contains recurrent hierarchical elements that can be decoded as instructions to the brain's operating system, the function of which is to rearrange the operating instructions in a manner that is conducive to the processing of particular input (i.e., to change the "default" settings, as noted earlier). In this regard, Tsunoda (1987) seems to have hit upon an important piece of the puzzle of how the brain might communicate with itself. Apparently, the Japanese language has within it a structural/functional unit that the brain interprets as an instruction to rearrange the pattern of hemispheric localization or activation (from

[4]Gedo's model can also be examined from a language perspective: "Mode I is prelinguistic, Mode II is organized around the protolanguages studied by Fónagy (see chapter 8), and Mode III concerns the lexicality of the natural language" (Gedo,1990, personal communication).

right to the left hemisphere) for certain sounds (a pattern that is not true for other Oriental cultures studied by Tsunoda).

It is anyone's guess what the nature of the linguistic code consists of. Two possibilities are suggested by Tsunoda's research. Since many Japanese words have onomatopoeic significance to Japanese listeners (my own impression, quite subjective, to be sure), one mechanism behind the unexpected left-sided localization phenomenon described by Tsunoda might be a linguistic reliance on crossed-sensory bridging, such as I observed nearly a decade ago playing a role in the insights following some psychoanalytic interpretations (which made special use of vivid metaphors). That is, the Japanese language may have become over years of evolution a special vehicle for the processing of certain kinds of (onomatopoeic) qualities in the auditory realm that have assumed a "logic" of their own (and therefore a left-sided localization).

Another possibility is that Japanese speakers first use the left hemisphere, rather than the right, to process the sounds mentioned earlier, because Japanese culture seems to place higher relative value (more perhaps than in the West) on decoding the "texture" of various emotional experiences, while at the same time not reacting overtly to the emotional content implicit or explicit within the message. Thus, it would make sense for the prefrontal cortex in Japanese speakers to attend selectively to voice sounds with the relatively emotionless syntactic power of the left hemisphere first. As is known, the right hemisphere has some language-processing ability but almost no ability in the syntactical area. Its forte is analyzing affect.

Beyond these speculations, however, we need to admit our ignorance. Only careful clinical observations of how the brain accomplishes its tasks will provide answers to the questions that we are raising regarding neural control and adaptive decision making. But, in reviewing the evidence, one is struck by the likely importance (in such adaptive processes) of the prefrontal cortex. Specifically, it seems likely that it is the prefrontal cortex (with its special ability with language and serial organization) that controls brain organization decisively, once natural (native) language is assimilated during development. In fact, the true importance of language, in an evolutionary sense, might be this ability it provides for adaptive, individual control over brain organization.[5]

---

[5]Some readers will object to my speculations on the ground that it is culture, and not biology, that is the decisive factor in the evolution of language. I think, however, this is an open question, (see chapter 11). It is also interesting to speculate whether the different languages (rather than one international language) evolved because there were selective advantages provided by the different languages, such as, for example, their adaptiveness in solving specific problems in the realm of different life contingencies (cultural contexts). For those interested in pursuing more about the subject of human language and cognition from an evolutionary perspective, I suggest Lieberman (1984). His central premise is that "human linguistic ability is based on rather general neural mechanisms that structure the cognitive behavior of human beings as well as that of other animals, plus a limited set of language-specific mechanisms that differentiate the particular manner in which we transmit information" (p. 1). My interest here is in expanding our understanding of exactly this relationship between neural control mechanisms and cognition, particularly the relationship between the language of the brain and that of the "mind" (see chapter 4).

6

Integrating Sleep and Dream Research

PRÉCIS

The late Max Stern wrote an important book, Repetition and Trauma: Toward a Teleonomic Theory of Psychoanalysis, *and I was highly privileged to write the introduction. Stern was interested in people and in what made them tick. His special area of interest, the focus of his book, was in bridging neuroscientific and psychoanalytic insights regarding the effects of traumatic experience. Henry Krystal's work in the area of trauma frequently builds on insights gained by Stern over years of carefully psychoanalyzing patients.*

It is of great interest to me that some of Stern's conclusions regarding the effects of traumatic states, and those of other schools of psychoanalysis (for example, conclusions of self psychologists regarding so-called arrests in development), dovetail neatly both with Freud's insights, as summarized in "Beyond the Pleasure Principle" (1920), and with pioneering research on sleep and dreams. The chapter that follows attempts to survey Stern's work on psychological trauma and then to carry it forward in a synthetic view of sleep and dreams. Although my theory of REM/nonREM sleep remains to be proven or disproven, it seems consistent with a large body of evidence within these two domains.

Because the details are important but too complex to review briefly in this précis, I would like to describe briefly just one of Stern's conclusions, the clinical significance of which will be obvious to the reader. The question is, Why do we have recurrent nightmares? Stern believed that such a pattern, especially when it reached the extreme form we call pavor nocturnus (that is, "night terrors"), reflects an arrest in psychological development. Stern was skillful in shifting back and forth between his knowledge of dreams per se, his knowledge of Freud, and his knowledge of the dreamer! The limitation (before Stern) was that all too often the researcher of dreams or sleep would pay too much attention to the phenomenon itself and not enough to studying the dreamer as a person. Stern combined humanism with scientific thoroughness, and hence he deserves credit for his synthesis (described in this chapter) of sleep and dream research, especially as regards the phenomenology of arrested states of psychological development secondary to early emotional trauma.

I

Quoting Valenstein and Schrödinger to the effect that the greatest of life's failures is not to try, Max Stern (1988) ambitiously calls for the reformulation of psychoanalytic theory so as to take into account the progress in biology in general and neurobiology in particular. A clinical psychoanalysis grounded in the best knowledge of the brain offers our field an unsurpassed pathway for growth; for psychoanalysis to ignore such developments, however, is to risk losing status with the public, stagnating as a science, or both. Other specialists, not as psychologically informed as psychoanalysts, would then be free to play increasingly important roles in health care delivery systems.

Stern lucidly reviews the growth of his reconceptualization of trauma from certain key clinical observations: of anxiety states and pavor nocturnus, of "traumatic" states, and of "regression" and "integration" within the psychoanalytic situation, where the fear of death is a prominent concern. At the core of Stern's effort is the challenge of integrating sleep research and dream research (see Gastaut and Broughton, 1965; Fisher, Byrne, Edwards, and Kahn, 1970). Reviewing these pioneering sleep laboratory studies, Stern uses their data to analyze the mystery of pavor nocturnus in adults and children. As I will attempt to show by introducing some of the work since Stern's seminal studies, we are gaining new clinical insights into how sleep and dreams may relate to each other, and to learning. In essence, we are learning how learning occurs, what disrupts it, and how to correct problems in learning acquisition.

The panic, dread, paralysis, confusion, and amnesia of pavor nocturnus resemble, for Stern, the catatanoid state of shock of central nervous system origin. Selye's classical work on shock points out how the brain is first aroused and then suppressed and even anesthetized by trauma. This effect is so substantial that surgery can be performed under its influence! It seems that the brain monitors the environment both outside and within the body with a view toward the early identification of threats, particularly those to the brain itself.

To assist the reader, let me add some general background. Kahn, Fisher, and Edwards (1978) show that pavor nocturnus is a relatively rare phenomenon in adults, more common in children, and with rare exception it occurs only during stage IV sleep (occasionally during stage III) just as do enuresis and somnambulism. Gastaut and Broughton's (1965) view, as noted, is that these night terrors are not triggered by (NREM) mentation, that is, that the mental content in pavor nocturnus attacks comes after the unpleasant arousal. In contrast, Kahn, Fisher, and Edwards (1978, p. 542) feel that a considerable amount of the vivid (NREM) dream content in pavor is indeed capturable, at least in some experimental subjects (more about this later). However, they are unable to confirm that the autonomic arousal events reliably precede the pavor attack itself. They are thus left with the puzzling conclusion that pavor nocturnus is preceded by a "physiological vacuum."

Stern concludes that pavor nocturnus represents "a defense against stress caused by threatening nightmares," a position challenged by some of Broughton's (1975) findings which Stern (1988) takes issue with and that seem to him disconfirmed by Fisher's later work (see pp. 54-55, esp. p. 52ff). He ultimately chooses two major foci for minute analysis: the agitation response and what he calls the catatonoid reaction, a form of catalepsy. The sequence in pavor is essentially that of a nightmare (in the preceding, stage IV sleep period) leading to an autonomic deficit, that is, the dreamer's reaction to his own thoughts is "an initial manifestation of shock" (p. 66). The nightmare, autonomic inhibition, and all the ensuing physiological shock reaction and defenses against this shock are then observed by the brain, are assigned meanings, and constitute the pavor nocturnus attack.

Vocalization occurs at the onset of the night terror, rather than at the end, where common sense might otherwise put it. It is this vocalization that the dreamer rapidly forgets. Rather than the scream or its mental contents leading to the physiological changes (heart pounding, sensation of rigid paralysis, immense pressure on the chest), the reverse seems true; these physiological changes are part of the shock and countershock that the brain is processing. That is, the brain both participates in and causes the (neurogenic component of the) shock reaction. The brain, in this sense, remains a not-so-silent witness to its own impending injury or even death, for along with the real (externally verifiable) scream, there is a kind of internal, paralyzed "scream," an emotional response to the potentially dangerous changes in pulse and respiratory rate.

The discussion in chapter two is, in my opinion, the most critical part of Stern's book. To clarify the discussion that follows, it will help to make some additional orienting remarks about pavor nocturnus and the unresolved controversies connected with it. To begin with, there is a tradition within psychoanalysis (and I believe Fisher's work, as noted, bears me out here) of seeing the dream state of pavor nocturnus as merely a special case of the regression that is generally exemplified by such things as dreams and symptoms. The core analytic idea is that during any such regression, the dreamer returns to previous points of fixation, presumably resulting from incompletely resolved conflicts at various developmental points. From this perspective the dream content (whatever is retrieved in pavor nocturnus) represents a by product of unconscious mental activity and can be analyzed as such (more about this later). What is controversial, however, regarding the special case of traumatic dreams is that no one yet knows for sure what their unique presentation signifies. For example, we are not sure to what extent mental content is really a factor in influencing what is traumatic about them. We are not absolutely sure either about the timing and significance of the autonomic changes that are associated with the phenomena (this is so especially because different investigators have obtained different data in this regard). Finally, we do not understand why some subjects have amnesia immediately for the mental content of the attack, whereas other do not (see Arkin, 1978, pp. 542-46, regarding these and other controversial points).

It may also help the reader to appreciate a second issue, related to but different from the issue of the dreamer's (or dream's) state of regression, namely, the dreamer's level along a developmental line. Stern pays attention to both issues when he considers dream and sleep phenomena as they concern pavor nocturnus. The point here is that not all dreamers are created equal: some have been arrested in their emotional development, and this fact in itself seems to be critical (usually undetermined in sleep research but more accessible in a psychoanalysis per se) in determining whether the product of their sleep will be nightmarish or not. Stern asserts, most cogently I believe, that "regression" during pavor nocturnus for the purpose of mastery over anxiety needs special qualification or elaboration if it is to be properly and fully understood. According to Stern, *the immature ego is incapable of dealing with severe conflict (trauma) without arresting its development.* He differs here decisively with Fisher, Byrne, Edwards, and Kahn (1970, p. 781), who suggest that stage IV nightmares are not dreams in the ordinary sense, but regressive phenomena with movements to earlier fixation points. Stern finds no evidence of "prearousal ego regression in stage IV sleep" dream material to explain sufficiently what happens clinically. According to Stern—who accepts Fisher's contrast between NREM and REM sleep and dreams—if Fisher were correct about regression, then experimental subjects waking up during stage IV nightmares would show some evidence of deeper or more primary process functioning, and subjects waking up during REM sleep should by comparison show shallower sleep. Actually the opposite is the case (pp. 71-72).

Rather than postulating a lack of ego regression, Stern posits "inadequate development" as the fundamental requirement for night terror attacks. Stern (1988) concludes at the end of chapter three that in night terrors *one is dealing with wishes arising ". . .in obedience to a compulsion to repeat, in the service of correcting a developmental failure to attribute meaning to a state of tension"* (p. 112). In the absence of proper assistance from the mother/caretaker, "dispositions will persist unchanged over time, as 'signals' of need to experience an external reality coordinated to an objective state of tension." It follows that *traumatic dreams are "not produced by psychic conflict." Rather, they can be traced back to arrests in psychological (ego) development resulting from a "lack of coordination with an original gratifying reality"* (p. 112, italics added).

It is significant that such conclusions, although identical with some conclusions of self psychology (Tolpin and Kohut, 1980; M. Tolpin, 1983; P. Tolpin, 1983), have been drawn by Stern from a different set of assumptions, in particular from a lifetime of work attempting to understand trauma in terms of traumatic states, traumatic dreams, shock and countershock mechanisms, and from an abiding interest in integrating the observations and theories of psychoanalysis with those of the neurosciences. That both self psychology and Stern's research efforts have in this particular area reached the same conclusion (from quite different starting points) suggests that the conclusion is fundamentally correct.

After rejecting Fisher's position on the question of prearousal regression in stage IV sleep in persons with pavor nocturnus attacks, Stern (1988) (pp. 76-77) agrees with Fisher's differentiation of REM versus NREM dream states: the "mental content of NREM sleep differs from that of REM sleep in being less lengthy, elaborate, bizarre, implausible, visual, and emotional, but more thoughtlike and conceptual in nature. . . ." Clearly, both Stern and Fisher recognize that during sleep something critical occurs within the brain, something that is reflected in dream and nondream states and that represents different aspects of cognitive/affective processing of emotionally meaningful experience. It is a tribute to the insight of both Stern and Fisher that very early they hit upon the importance of the REM/NREM distinction as relating to fundamentally different ways of processing information, a conception that has bridged sleep and dream research, just as it has stood the test of time.

Stern (1988) then takes up the subject of trauma and the repetition compulsion. Night terror attacks "are. . .repetitive efforts to establish a beginning coordination between a state of tension and external reality. . ."

(p. 113). He summarizes with a tidy dichotomy: NREM sleep "is devoted to the resolution of the . . . problem [of] attribution of meaning to one's own states of tension," which Stern sees as the central problem Freud (1920) tries to solve and explain in "Beyond the Pleasure Principle"; REM phases "in contrast, would be concerned with the resolution of problems involving ambivalence, which would account for the greater appearance of primary process distortion in dream reports elicited from them" (p. 113). That is, for purposes of "defense" some distortion is expectable. In Stern's view, NREM ideation serves the function of mastery, and the repetition compulsion represents "mental activity more primitive and elementary than that associated with conflict" (p. 114). By this neat parsing of function, Stern has enabled us to imagine a psychological division of labor within the brain that fits with some of what is known about the organization of episodic versus semantic memory[1] between the two hemispheres (see Basch, 1983).

The final two chapters of Stern's book deal with a unitary conception in the form of a "teleonomic principle" of biologic activity, a derivative of the thinking of the geneticist Monod. Human evolution is teleonomic, that is, goal directed, meaning that the genetic blueprint is paramount; thus all knowledge is acquired through learning programs that are themselves genetically determined and, one might add, species specific. From the perspective of learning, Stern might appear to be asserting that trauma asserts its effect primarily by interfering with the learning that ordinarily would have followed a genetically preprogrammed pattern, but that instead is aborted. For example, he describes a situation where traumatically induced frustration during infancy or childhood leads to a disturbance in the "pleasure self" (p. 129) and to magical attempts to undo old frustrations.

Clearly, Stern does not make exactly the assertion just mentioned. I

[1]In Chapter 2 I described the critical difference between "episodic" and "semantic" memory processes, the former based on personal memory and affectively charged feedback cycles, and the latter coinciding more with general knowledge or deductive thinking. My point was to relate these psychological categories to the characteristic qualities of the two cerebral hemispheres, a line of thinking, supported as well by Basch (1983), that suggested to me that repression and disavowal might be defined as interhemispheric communication blocks in different directions. In a similar manner, if one keeps these distinctions and possibilities in mind, one can see that Stern's clinical theorizing leads inexorably to a view of dreaming as representing selective activity within various learning subsystems of the brain, including, of course, roles for each of the two cerebral hemispheres in REM and NREM dreaming. Later in this chapter these points should become substantially clearer to the reader as they are elaborated.

speculate that he "might appear to be asserting" such a generalization regarding trauma. In fact, I myself am raising this possibility, although I believe Stern's view is nearly the same. The justification for this hypothesis is that it focuses psychopathology upon a learning pathway that is vulnerable to biopsychosocial disturbances. In so doing, of course, I am not contesting the clinical usefulness of recent psychoanalytic studies of trauma, but only suggesting the usefulness (see for example Rothstein, 1986) of a novel, interdiscplinary approach (relying, for example, on learning theory), especially when one is otherwise left with a collection of rather different and possibly fundamentally unintegratable approaches. Although Rothstein's (1986) efforts in this area are exemplary, for example, they do not really solve a difficult clinical problem that requires some overarching theory.

To continue with Stern's reasoning, the reenactments that are at the core of the repetition compulsion occur under the primary effect of experience in the here and now. From my perspective, this essentially releases the "procedural" memories (versus declarative memories)[2] of the earliest (disturbed) sensorimotor schemata (see chapter 3). Stern (1988) himself refers to this as "the emergence of...the primarily repressed" (pp. 129-130).

One may ask if Stern is correct in endorsing Monod's thinking of a "teleonomic principle." My own sense is that to understand the human brain requires special consideration from a number of perspectives. For one, we have amazing brain plasticity,[3] even compared with related

---

[2]Within psychology "declarative and procedural" memory systems have been described (see Squire, Cohen, and Nadel, 1982, and Squire, 1986). The former represent those memories which are best exemplified by learning at school. These would include learned rules, such as those for grammar or mathematics. Such "declarative" knowledge is usually retrievable merely by questioning the individual about particular rules. In contrast, "procedural" memories are not available to questioning, but need to be primed by experience within some related sensorimotor activity. This is more like associating within an associative network; but entry into this kind of memory requires an experience, and what is retrieved is a display of the knowledge involved, which is proof that it was in memory in the first place. Such dichotomous memories clearly point to the existence of at least two, quite different learning-related systems. Actually, of course, the brain is made up of many such systems and this subject is taken up in my previous efforts, as well as in the second part of this essay.

[3]By "plasticity" I am referring to the capacity to learn and grow emotionally, as reflected by the modifiability of the brain's anatomy (cytoarchetectonic details) or altered functioning of the modular systems of the cortex (Szentagothai, 1975) in response to experience. This subject is more fully discussed in chapter 3 and is of

species. For another, we alone in the animal world have developed a material culture that contributes to our learning in ways that might not be predictable if one were to look exclusively from a biological perspective. The particular point of Monod's with which one might quarrel has to do with our being able to alter our genetic programming. Monod believes that this sets severe limitations; I am not so sure. Although Monod cannot be proven wrong categorically on this point, given the known capacity of the genome to alter its expressivity (McClintock, 1984) and our clinician's sense of human changeability, Monod's point seems exaggerated. Where one can agree wholeheartedly with Stern, however, is that given human neotony "the need for external assistance in the face of disorganizing states of tension" (p. 147) is absolutely essential. In this sense Stern seems correct in quoting Monod to the effect that a teleonomic principle does seem at work: trauma leads the human species to learn adapted behaviors that provide people, at times of psychological stress, with the necessary signaling and learning systems for obtaining help from others.

Stern, then, supports the concept of what has been referred to as selfobject functions (Kohut, 1971). He writes of the patient's state of tension, which is essentially mastered by someone other than the patient himself. For Stern no human is spared biotrauma in this sense, and our failure to master this "is at the very heart of mental disorder" (p. 118).

Finally, Stern considers specific analytic case material, especially regarding perversion, feminine masochism, and compulsive personalities, all under the rubric of conflict psychology, as opposed to self psychology. For "fragmentation" anxiety, he substitutes the fear of death; in place of "grandiosity" of an archaic self, he refers to "demands for a magic formula" and "associations with Christ." This clinical material is written in an entirely unpolemical way, with clarity, intelligence, and compassion. One gains the impression of a seasoned and innovative analyst who has gone to unusual lengths to help traumatized analysands uncover new psychological capacities. For example, he helps these patients learn to use their own drawings and listen to tapes of their analytic hours to enhance their self-reflectiveness and introspection and to prime their memories.

The interested reader may well wish to spend time with Stern's work, written over a number of decades. A review of Max Stern's work makes it obvious that he never hesitated to make bold leaps in attempting to explain difficult phenomena, such as the nearly universal fear that death holds for man. Stern returned over and over to this question with the insights regarding trauma and structuralization that I have discussed.

---

critical importance in understanding learning, including learning in the psychoanalytic situation.

Extending out from his consideration of pavor nocturnus, an early form of death angst that seems to be not uncommon in children although fairly rare in adults, are a number of related subjects that form a vital area of research today: panic attacks, phobias, catalepsy, shock and autonomic arousal, posttraumatic stress syndrome, and the general subject of biotrauma. Of course, Stern's major avenue of approach to this research was via his interest in sleep research and the closely connected subject of psychoanalytic dream research.

It has long been a goal of researchers to bridge sleep and dream research. There seem to be a number of paths along which such bridging appears to be possible. On the biological side, we now have (1) a variety of REM/NREM studies,[4] including developmental and evolutionary perspectives; (2) research on basic rest/activity cycles, circadian and ultradian biological rhythms, and entrainment phenomena; (3) studies correlating regional cerebral blood flow with a host of psychological variables and using a variety of noninvasive techniques; (4) research on brain stem mechanisms relating to pontine mechanisms or brain stem "gating" phenomena, and other systems aspects between REM and REM-related changes in the brain stem and in other systems such as the forebrain, limbic system, and the like. On the psychological/psychoanalytic side are (1) work on dreaming in general; (2) research on "self-state" dreams in particular; (3) studies of the relationship between dream/sleep state, learning, and memory; and (4) studies of sleep/dream-related learning facilitation and inhibition (sometimes referred to as REM deprivation studies). These latter learning studies are extremely difficult to categorize

[4]Throughout this essay, the dichotomy of REM versus NREM sleep is taken to be something the reader already understands. A few words may help some readers confirm the details of this understanding. Aserinsky and Kleitman (1953) discovered the REM, or rapid eye movement, phenomenon in man, which seems to attend most but, importantly, not all of our dreaming. REM and NREM periods are specific phases of the sleep process. Sleep has been characterized (Tobler, 1984) behaviorally, electrophysiologically, and physiologically as follows: Behaviorally, sleep has a typical body posture, physical quiescence, an elevated threshold for arousal, rapid rate reversibility, and circadian rest-activity cycles; electrophysiologically, the electrooculogram shows periods of rapid eye movement five or six times each night (so-called REM, or "paradoxical," sleep), followed by periods during which the depth of sleep rises and falls (so-called NREM sleep, stages I through IV); the electromyogram shows progressive loss of muscletone, which loss becomes maximal during REM periods; and the EEG or brain wave shows low voltage, fast waves during REM periods, and high voltage, slow wave spindles during NREM periods. Finally, physiologically, throughout sleep there are periodic fluctuations in heart rate, respiration, body temperature, genital tumescence, and so forth.

since they have been conducted under an array of orientations: some conceptualize in terms of left-right hemispheric brain mechanisms, some in terms of the concept of brain "plasticity," and some from the perspective of the changing brain organization of large subsystems of the brain as a likely concomitant of so-called psychological development.

Given the complexity of the subject and the large number of possible approaches, I would now like to review some of what I consider to be useful work in the field of sleep dream research, that is, work that seems to me to allow some significant bridging to occur. In doing so, I will try to be relevant and clarifying rather than comprehensive. In general, I will follow the major biological and psychological areas noted earlier.

I wish to state at the outset that Max Stern's work is the inspiration for the comments that follow.

II

Freud's (1900) "Interpretation of Dreams" was the beginning of scientific dream analysis. Subsequent psychoanalytic dream research of note (other than those research efforts mentioned in the first part of this essay) includes the contributions of French (1952) and Fromm (1947), who focused and extended dream interpretation within the Freudian tradition; Altman (1975), who summarizes Freud's lengthy and detailed dream treatise into one highly readable volume; Friedman and Fisher (1967), who relate basic rest and activities cycles of the brain to dreams; and Berger (1967), Greenberg (1970), Wasserman (1984), Gabel (1985), and Slap and Trunnell (1987), who have each made special clarifications regarding the task of bridging dream with sleep research.

The position of Freud regarding dreams, simply stated, is that when impulses and wishes from early, instinctual life (latent dream thoughts), and current experience (the day residue) resonate sufficiently with and reinforce each other, then one's latent dream thoughts are converted into a manifest dream by way of process Freud referred to as dream work. This work involves a variety of mechanisms, including condensation, displacement, symbolization, and pictorial metaphor formation. Once produced, the hidden meanings of the dream continue to be protected from discovery by the continuation of defensive mechanisms. These include forgetting part or all of the dream, forgetting dreaming itself, secondary revision of the dream at the time it is remembered or retold, and even conscious withholding in relating the dream to others. The early stages of psychoanalysis were directed toward reversing these processes and getting at the unconscious sexual, aggressive, and other wishes thereby exposed.

Gabel (1985) contrasts Freud's view of dreams, for what they conceal, with Jung's view of dreams, for what they reveal of purposive, adaptive, and even "prospective" cognitive patterns. Gabel's inspiration comes from Montague Ullman (see Gabel, pp. 190-191), who, along with researchers such as Greenberg (1970), Berger (1967), and others, has formulated theories that seek to explain how dreams relate to such specific brain functions as sensory perception, the coding of experience, and the storage, retrieval, and organization of memory.

Before I continue, however, I would like to note some important issues. One is the problem of dealing with the information-processing aspects of dreams without shifting the focus so as to destroy the meaning of dreams as described by Freud, which has proven monumentally important in clinical psychoanalysis. In addition, we also do not want to gloss over the differences between Freud and Jung, which are considerable and which add another problem of deciding how to consider the possible meanings of any particular dream. Finally, in any synthesis we want to bridge the humanistic and scientific traditions represented within each relevant theoretical approach.

To begin, it is clear that Freud unequivocally rejected Jung's view on dreams. For example, Freud (1916/1917) writes of some psychoanalysts who erroneously assert

> that dreams are concerned with attempts at adaptation to present conditions and with attempts at solving future problems—that they have a "prospective purpose". . . . We have already shown [however] that this assertion is based upon a confusion between the [manifest] dream and the latent dream thoughts and is therefore based on disregarding the dreamwork. As a characterization of the unconscious intellectual activity of which the latent dream thoughts form part, it is on the one hand no novelty and on the other not exhaustive, since unconscious intellectual activity is occupied with many other things besides preparing for the future [pp. 236-237].

Further, Jung seems no less to have rejected Freud's views on dreams. In particular, Jung rejected Freud's sense of the critical importance that dreams hold for gaining access to hidden sexual and aggressive impulses, which according to Freud is at the core of dreams as well as the psychopathology of everyday life. To Jung, sexual and aggressive motives are part of life, but substantially less important than such other issues as "individuation" (Henderson and Wheelwright, 1974, p. 817) and "the inherited structure of the brain," which Jung referred to as the "collective unconscious" (p. 809).

Because of this historic controversy it is possible to read various statements that seek to bridge dream and sleep research as fundamentally continuing the controversy in some polemical manner. This is something I wish to avoid, as much as is possible. I hope this can be accomplished by focusing more upon dream and sleep phenomena per se and as they relate to each other and less on what is implied about fundamental human nature.

Beginning with some of the early work on dreams and sleep, Greenberg (1970) states that

> the dream process serves to bring together perceptions of recent emotionally meaningful experiences with memories of past experiences of a similar nature . . . the new experiences can be dealt with in the same manner as the earlier experiences, or the new experiences might indicate that the earlier experiences no longer need to be handled with the old, characteristic but outmoded methods of adaptation. This latter event can be seen as effecting a change in the memory system rather than just adding new information to it [p. 265].

Greenberg credits Dewan with the idea that memory filing seems most likely to be "according to emotional tone" (p. 265; also see Dewan, 1970). Greenberg also cites still earlier research by Pariagiani and Zannoco in 1963 and Passouant and Cadhilac in 1962 for showing that "hippocampal activation [occurs] during paradoxical [i.e.REM] sleep in cats" (pp. 259-260). Importantly, this kind of hippocampal theta activation has been related by Meissner (1966) to learning and memory processing in animals.

In his early work, Berger (1970) clarifies that REM sleep serves multiple functions, a finding which the literature continues to confirm. These functions include maintaining "the integrity of the CNS process involved in the coordination of eye movement, which coordination would be lost if there were extended periods of sleep without periodic [REM sleep] enervation of the oculomotor system" (p. 278). According to Berger, Dewan feels that the key in all of this is that the REM "reprogramming" occurs offline, that is, during sleep (p. 296). It thus gives the organism a chance to gain in efficiency during the day by saving information for processing later on, thus relieving the organism of the need to both perceive and process information simultaneously. This idea is later reintroduced by Winson (1985) in reviewing the subject of dream and sleep research.

One cannot review this research without commenting, if only briefly, on the technologies involved. Several methodologies have provided information about selective information processing in man. Since the 1960s it has been known that the brain generates electrical potentials in relation to high level cognitive operations (Desmedt, 1979). Unfortunately, "the problem involved in obtaining valid correlations between [such variables as] ERP [evoked response potential] components and selective processes in man are formidable" (Hillyard, 1979, p. 9); also, the focus until now has been to study subjects who are awake, where the cognitive task and evoked potential response can more easily be related to each other. There apparently is very little that exploits this technique to understand fundamentally the information system activity during dreaming. In addition, from approximately the same time period, regional cerebral blood flow studies (see Lassen, 1987, for a review of research in this field; also Roland and Friberg, 1985, and Friberg and Roland, 1987) have provided further clues regarding complex brain information processing during the waking state. Again, to my knowledge nothing has been reported on sleeping subjects.

What is important here, however, is that unlike the electrical studies, CT scans, and MRI techniques that are available, the radioactive xenon scanning techniques of Lassen and other researchers for following regional cerebral blood flow seem potentially the most valuable for the ultimate bridging of sleep and dream research. This is because they are supremely functional, that is, they show us what the brain is doing. (PET scanning is too expensive and too infrequently done to be of much practical value). It is unfortunate that we have already spent scarce funds for such expensive equipment when it is not likely to help us learn what is going on (for example, during sleep) in the deeper structures of the brain. When more xenon scanning machines are available (of the latest type developed by Lassen) and studies have been extended to sleeping subjects, we shall have our first detailed insights into what the processing patterns are between and within deep and surface brain structures during the sleeping state. We will then be able to test out some of the hypotheses that attempt to integrate dream and sleep phenomena.

But there is another reason for the lack of progress in bridging dream and sleep research. Unfortunately, while psychoanalysts and others have been trying to understand the research on REM sleep, a number of brain scientists, not satisfied with simply making discoveries in their own field, have chosen to criticize psychoanalytic dream theory. This has led to articles like Wasserman's (1984) and Rechtschaffen's (1983), which point out the faults in such criticism as that from Hobson and McCarley (1977). The problem is that this effort to defend psychoanalysis from unwarranted criticism has required time and effort that would better have been spent in reviewing REM research findings, evaluating the evidence, and integrating insights with what is being accomplished within psychoanalytically oriented dream research.

A third reason for the failure in effecting the bridge is the failure of psychoanalysis to appreciate some of its own dream research. An important area of this research has been so-called self-state dreams, a designation attributed falsely to Kohut in a psychoanalytic article by Slap and

Trunnell (1987). In fact, self-state dreams seem to have been part of Jung's contribution to psychiatry (see Gabel, 1985), and although Kohut and others (Tolpin and Kohut, 1980; P. Tolpin, 1983; M. Tolpin, 1983) deserve credit for rescuing this insight, I do not believe they initiated it. Moreover, although Kohut did discuss such dreams, he did not introduce the subject in 1977, as indicated by Slap and Trunnell (p. 252), but in 1971, in *The Analysis of the Self.* My intention here, however, is not to find fault with scholarly psychoanalytic articles. Rather, it is to rescue the important concept of self-state dreams, which might otherwise be buried within self psychology and forgotten. There is an unfortunate tendency to praise overmuch certain theories when they are new and to discard their hard-won insights when they become subject to criticism. I think this is exactly what has happened with self-state dreams. In what follows, however, I will try to demonstrate that Jung's insights, along with Kohut's (and the Tolpins') *regarding this aspect of dream interpretation* are extremely important to clinical practice and that they are not necessarily at variance with the sound practice of psychoanalysis according to a conflict model. There is no reason, in other words, why dreams cannot be examined from both conflict and nonconflict perspectives, just as Max Stern has done, to see which is more apt in a particular case or whether or not both views might prevail. In fact, we have in psychoanalysis ample precedent for doing exactly this in Freud's use of the genetic, adaptive, topographic, and other psychoanalytic viewpoints, more or less simultaneously.

Particularly interesting about the self-state dream, in my opinion, is that there is now an abundance of information in neuroscience that relates to how experience is perceived, stored, organized, and retrieved; and this new scientific study of learning as a process has a very great bearing on the learning blocks that occur in persons who identify themselves as in need of psychiatric help (see chapter 3). In fact, *it may be the recognition by the experiencing self of the very existence of those learning blocks and the dangers they represent that self-state dreams announce.* And when such dangers pass some critical threshold, so that external help is needed, then we have a nightmare instead of a dream. If these speculations prove correct, then we must make use of all that we know psychoanalytically and neurologically to better understand our analysands. This would mean properly decoding the self-related danger in such traumatic dreams.

My thinking about this integration (and what I am about to speculate about) has crystalized over years and has grown from the daily attempt to understand dreams with my patients, either in psychotherapy or psychoanalysis. It also flows from a lifelong interest in the brain and how it works, although my formal research in integrating psychoanalytic and neuroscience perspectives really started in 1971, when I began my residency in psychiatry under Roy Grinker, Sr., who will, for me always be the best example of a disciplined scholar of the mind and brain. During my training I also had the opportunity to study with Michael Basch and to meet briefly but significantly with Ludwig von Bertalanffy, the father of general systems theory. Both had a profound effect upon my thinking.

I would like to try to describe now the synthesis I have made, leaving the bulk of the details for the third section of this introduction. This will make it easier for the reader to understand what follows, as well as making it easier for me to direct my own thinking in a highly complex area. My thinking begins with the following ideas. Sleeping is not really necessary for life but is an expression of an evolutionary trend in brain development, that is, of the particular pathway through which our particular kind of (human) brain came into being (see Tobler, 1984). In the same way, even the other life forms that sleep do not all have REM/NREM cycles built into them. Thus, some animals do not sleep, and others sleep in a different way than we do. For example, low voltage, fast-wave sleep (i.e., REM sleep) is absent in fish and amphibians, questionable in reptiles, and definitely present in birds and marsupials (p. 215). Yet if one looks at the EEG for differences between vigilance states, one can definitely find these present in reptiles and even in some fish. Thus, as we move up the evolutionary ladder, we come to life forms that sleep with REM and NREM periods as an important pattern whose brains are organized in this way for a purpose that is both discoverable and adaptive. I feel that the person who has come closest to discovering this purpose is Dewan (1970) who speculated about the efficiency of offline processing. I think we can now specify even more about exactly what this means.

My main points concern what specifically the organizational changes consist of when, during alternating periods of REM and NREM sleep, something is occurring that is critical for learning. My impression is that the organizational changes occurring in the brain during sleep relate to the following specific system relationships and confer some adaptive value, such as planning for biopsychosocial realities: (1) There are at least three major learning-adapted systems of the brain, and these, during REM periods, repeatedly "review" the day's residue, which they have individually received during the subject's recent experiencing. This review is possible because, during the REM periods, critical learning systems or their parts are relatively separated from each other (see Bakan, 1978). By "review" I mean the creation of a long-term memory storage form for particular memories, which were previously in some intermediate form. I also assume that a critical step in this process involves some final rearrangements in the organization of these memories, for example, according to affect (see Dewan, 1970). (2) These three systems are the vestibulocerebellar system (VCS), the corticostriatal system (CSS), and the corticolimbic system (CLS), each of which could be related to a particular category of knowledge and orientation, which I will discuss presently. (3) Each of these three systems, in my opinion, has its own vocabulary and unique "perspective," which differ to some degree from the others. One of the next necessary tasks in information processing might be the communication of these three great systems with each other, downloading, which I believe occurs primarily during the NREM periods (when these major information-processing systems are connected). That is, each of these three systems is outputting and inputting its load of information into or from the other two systems during NREM periods. This downloading must result in further learning for each system, since each receiving system will have loading for the same "event" but from no less than three quite different "perspectives." I further assume that these processes of review and downloading are connected to each other, or interdependent, and that this explains the need for alternating REM with NREM periods. (4) The effect of this sharing of information evidently is to update the primary database within each system, and with this, each system probably can begin to have some emotional or other reaction to this exchange. Specifically, the potential reactions seem likely to include as well the patterns earlier described by Stern for REM and NREM sleep. (5) In my opinion, the system relationships between these three major learning systems are critical. It is also possible that the smoothness or lack of integration or cohesion within the overall operating system that we call brain is very much effected by how well these three subsystems work with each other. In a parallel way, I have discussed in chapter 2 how the two cerebral hemispheres can either collaborate or conflict, with potentially serious effects on the overall personality functioning. (6) This means that we can consider the dream not only as a reflection of the CLS, as I believe Freud did, in what we call the conflict-psychological perspective, but as reflective of nonconflict-based issues, indicating, I believe, the functioning of the two other systems as well, the CSS and the VCS. To my mind, these two additional systems are primarily concerned with the status of the prerepresentational self (see Emde, 1983) and are designed to accomplish a different set of tasks than the CLS. In fact, what have been called self-state dreams, may be reports of the status of these latter two systems, particularly when these systems register a condition of danger to the self (as occurs in nightmares, but is not restricted to such traumatic dreams).

I have considerably extended the number of assertions and questions that must be answered. At least, however, I have laid the ground work for filling in some of the details; I have also introduced the possibility that these issues of brain function (learning subsystems) and dream state might profitably be related to each other. And I have given us some clues to how we might proceed with the necessary clarifications.

One clarification that can be made at the outset concerns what I meant when I said that each of the three major learning systems is connected with a particular traditional domain or viewpoint. Let me explain. The CLS represents the traditional psychoanalytic perspective in particular, but also the psychological sphere in general, since it is concerned with integrating the great limbic system (and therefore our right-hemispheric, affect-regulation system) and that which we learn through society (the left-hemispheric language and cultural, rule-related system). In psychoanalysis, as it was begun by its founder, the issue was of integrating these two opposing hemispheric perspectives that placed the person with sexual or aggressive instincts/drives in the problem situation of dealing with particular cultural prohibitions.

In contrast to the CLS (or CLDS, for the corticolimbic diencephalic system), the VCS represents the leading edge of the biological sphere: it relates more to the establishment of a core sense of self, or the self-in-the-world model (see chapter 3). Frick (1982), in a particularly valuable contribution, relates the development of this particular system to the establishment of the ego (or, one might say, the self). To my mind this learning system, available since birth (see chapter 10), is the biological basis for the establishment of the prerepresentational self (see Emde, 1983). Trevarthen (1979, 1985) has also pointed out the complexities involved in specifying which neuroanatomical structures might be involved in developmental steps; and although he might disagree with the particulars suggested herein, I think he would agree with the spirit of this enterprise, namely, that it is worthwhile to begin to think about what brain systems and what behavior are related to each other. The VCS, then, explains some of the functions observed in early child observation studies, whether psychological or preoedipally psychoanalytic.

The third system, the CSS, represents the leading edge of the social sphere insofar as it is an excellent system for acquiring habit patterns, and these are the building blocks of social roles. This system has been ably described by Mishkin, Malamut, and Bachevalier (1984) and refers, again, to a system that (in contrast to the CLS, which requires years for its maturation and numerous experiences for its registry) is available from birth and often requires only a single sensory experience for the learning to occur! Young children use the CSS function as competently as adults. This system explains the data accumulated over years by those in academic psychology and other specialities that relate to the acquisition of habits (versus more complex memories, such as are acquired through discriminative learning).

Of course, another obvious reason for paying attention to the differences in the functioning of these three different systems is that armed with this perspective one can avoid getting caught up in needless debates over what learning is. For example, as Mishkin et al. state: "If both habits and memories are constantly being formed by experience in normal animals, then the great debate between behaviorists and cognitivists will have finally been resolved in favor of both parties" (p. 73). That is, we (like many simpler life forms) not only employ our VCS and CSS to adapt, but we also simultaneously tap our CLS and acquire knowledge in a cognitive fashion. So learning is quite a variety of things. These include all the ways in which experience is capturable by the plasticity of the brain (see chapter 3), as well the changes (on a large scale, or system level) of organization of the brain. And we aspire eventually to be able to correlate psychological development with such changes within the brain (see chapter 10). Another reason for paying attention to the system source of learning is that if we are to ever fully understand our patients and their learning blocks we must learn to recognize the "fingerprints" of the major learning subsystems of the brain involved.

III

Research by Galin (1974) and Broughton (1975) supports the correlation noted in the preceding section between REM sleep, dreaming, and activation of the right cerebral hemisphere. Bakan (1978) seems to have been the first to suggest that during REM sleep periods the right hemisphere is released from left-hemisphere control, a point I have used in my own speculations regarding the effects of this release on "review" and "downloading" activities of important learning subsystems (see the previous section). Flor-Henry (1983) presents data supporting the corollary view to Bakan's position, namely, that during development the left hemisphere asserts control (dominance) over the right hemisphere during waking activity and that this is part of what prevents the acting out of forbidden sexual and aggressive impulses. Of course, during sleep the potential for acting out is substantially reduced.

Bertini (1982) shows that subjects differ widely in their use of sleep and waking states in information processing. Specifically, he notes that whereas there are some persons "for whom dream mentation is strictly associated to REM [periods]," there are others "for whom this association is much less strict or rigid." (p. 59) Bertini, using a unilateral tactile recognition task, concludes that *there is in general "a right hemisphere dominance during the REM state"* (p. 59, italics added). He further notes that "people who exhibit strong lateralization during wakefulness are also the people who show strong REM specialization" (p. 59).

This kind of correlation, supported by Galin's (1974) and Broughton's (1975) work, is appealing, because it fits with what is already known about the right hemisphere's special capacities for gestalt formation, spatial perception, and management of affect (through its preferential connection with the limbic system). However, before one can conclude that REM dreaming is led by a system primarily determined by the right hemisphere, one needs to review allegedly contradictory evidence. In this regard, Antrobus, Ehrlichman, Wiener, and Wohlman (1982), using EEG monitoring, assert that "visual imagery is [actually] stronger as the left, not the right, hemisphere becomes dominant!" (p. 51). Accordingly, they "do not find support for the general arousal or state dependent models with respect to . . . EEG activation [of the right hemisphere]" (p. 51). In addition, Lavie and Tzischinsky (1986), investigating the relationship between cognitive laterality and REM sleep, conclude that "dreaming cannot be seen as a right hemisphere function" (p. 353). Of course, the claim being made is not that all dreaming is right hemispheric, but rather that REM dream states seem to be a reflection of the right hemisphere's activity, at least in certain individuals. Reading the fine print, however, one is less convinced about the arguments of Lavie and Tzischinsky. For one, they did confirm that as far as right-handers (but not left-handers) are concerned that subjects awakened during REM periods did better on tests of right-hemisphere function (p. 355). Moreover, left-hemisphere kinds of testing also discriminated when right-handed subjects were awakened during NREM periods (pp. 355-357). Rather than disproving the hypothesis being questioned, these findings seem to qualify it, much as Crow's (1986) work on schizophrenic twin studies in England has shown that there is concordance between schizophrenia and temporal lobe abnormalities in right-handed identical twins, but not left-handed identical twins. It seems to me that a reasonable conclusion is that REM dreams and right hemisphere activation are probably closely correlated, at least in right-handed people. This should not be taken to mean, however, that dreaming is a right-hemisphere phenomenon; but rather that the right hemisphere seems to be a leading edge in the REM type of dreaming, so that its "release" is an important element in the information processing and exchange between (learning) subsystems of the brain. It is my feeling that understanding the organization of such subsystems ought to be our highest priority. This was the goal when Vuckovich and I (chapter 2) reviewed interhemispheric communication from a psychoanalytic perspective; and this is the critical part of our work on brain plasticity and learning, where we consider the possible role of the cerebellum and its nuclei in bridging the hemispheres and thus coordinating some of the brain's learning activity (chapter 3).

Finally, an article by Ehrlichman, Antrobus, and Weiner (1985) needs to be mentioned to complete the review of the research critical of the right hemisphere activation/REM dream hypothesis. One problem in this research is that the authors use an EEG power asymmetry measurement technique developed for studying waking subjects and then apply it to identification of hemispheric activation in sleeping subjects, a task whose certainty they themselves recognize as difficult (p. 482). A second problem lies in their extraordinary idea that "dreams are not primarily visual phenomena" (p. 483), an assumption that seems questionable in the extreme. On top of this, they then contradict themselves by suggesting that "the visual aspect of dreams cannot be denied" (p. 483), changing their minds primarily because they now wish to argue that REM sleep dreams cannot possibly be right hemispheric because some data suggest to them that the visual imagery system involves in part some left-hemispheric functions. Finally, they make the unlikely inference that "mentation in REM [sleep] is continuous with that in waking [life] and will thus show the same patterns of hemispheric involvement as would comparable cognitive activities in waking" (p. 483).[5] This latter assertion, I believe, is contradicted by a large mass of sleep and dream research that establishes beyond doubt that what is happening during sleep and dreaming states is quite different in its organization and impact from waking mentation, such that disturbing these sleep/dream (e.g. REM) states clearly has an impact on learning acquisition. In short, Ehrlichman and his colleagues fail to be convincing, and even if they are correct in some particulars their work seems off the main line.

At this point, I wish to move from dreams and communication within and between brain subsystems to the subject of (REM) dreams and learning per se. W. C. Stern (1970) shows with early REM deprivation studies that there is a clear correlation in humans between REM deprivation (RD) and learning impairment. This subject is also reviewed by Smith (1985), who covers animal research, which finds the same clear correlation between RD and learning impairment. W. C. Stern's work further quotes studies by Dewan and by Gardner to the effect that one can

[5]In the discussion of the transference phenomenon itself (chapter 8), I make a point about the similarity between REM activities, dreamlike states, and the transference phenomenon. Yet in the foregoing passage, I criticize Ehrlichman et al. for seeing REM sleep mentation as continuous with waking states. His may seem to be inconsistent reasoning, but actually it is logical : I am criticizing Ehrlichman et al. for missing the point that REM seems to be a special cognitive state (with right-hemispheric qualities), quite different from the ordinary (left-hemispheric quality) cognitive activity of NREM sleep. But when I later claim, that REM and transference states appear to have a close correspondence, I am merely asserting that REM sleep cognition and wakeful transference cognition are also strikingly similar (see also fn3, chapter 8).

conclude without much doubt that *"new memories are [primarily] processed and encoded during REM sleep"* (p. 255), that is, during our REM dreams. Dushenko and Sterman (1984) present further experimental support for Bakan's theory that REM *sleep deprivation works against learning primarily by its impact on the right hemisphere, which is in "cyclic ascendence" during REM sleep* (p. 25). *That is, REM and NREM cycles represent cyclic alterations in the connectedness and communications between the two cerebral hemispheres, and by inference, between the leading information processing/learning subsystems of the brain noted earlier in this chapter.* (For additional discussion of some of these issues, which are beyond the scope of this brief essay, the reader is referred to Klein and Armitrage, 1979; see also chapters 3 and 10).

To round out this discussion of sleep and dream research, let us consider some basics regarding the circadian sleep wake cycle and its regulation. Gross (1982) reviews this subject in depth, noting that the basic pacemaker in humans is set for 25 hours and is probably controlled by an oscillator in the hypothalamus (p. 21), itself a subsystem of the brain of extraordinary complexity. Entrainment phenomena are mediated visually by a retinohypothalamic projection to the suprachiasmic nuclei (SCN) (p. 21). A number of substances, including vasoactive intestinal polypeptide found in high concentrations in SCN cells, probably play a key role in inducing sleep (p. 26). Borbely (1986) discusses research on endogenous sleep substances. He basically conceptualizes sleep, however, as a product of a sleep-dependent process (process S) and a circadian oscillator (process C). How these processes are connected pharmacologically, and how they relate to dreaming, are not specifiable, except as reviewed in the first parts of this essay. Jouvet (1982) pursues 5 hydroxytryptophan (and other indolamines), which plays a critical but as yet incompletely understood role, along with peptides and other "hypnogenic factors" (p. 93) in controlling sleep rhythms. Sakai (1984) adds that two different populations of cells in the medial part of the nucleus reticularis magnocellularis "are closely tied to transitions into and out of paradoxical [i.e. REM] sleep" (p. 4). Put another way, Sakai is telling us that pontogeniculo-occipital (PGO) waves "located in the caudal mesencephalic and rostral tegmental structures" (p. 5) of the pons are associated with vivid dreaming in man and other mammals. Most important for psychoanalysis, and contrary to the views of Hobson and McCarley (1977), the localization of this PGO system does not explain dreaming, nor does it in any way invalidate psychoanalytic insights into dreams, which have been obtained by a unique methodology, namely, clinical psychoanalytic experience. What this research does clarify is that the postural atonia occurring during REM periods is secondary to the "tonic excitation of supraspinal inhibitory system" (p. 10) associated with PGO activity. And the ascending reticular system originally described by Moruzzi and Magoun (1949) "also plays an important role in the mechanism underlying cortical desynchronization during PS [paradoxical sleep, which is the same as REM sleep]" (Sakai, p. 14).

A third complication is that hypothalamic-hypophyseal control should not be ignored in trying to understand these phenomena of sleep and dreams. Makara, Palkovits, and Szentagothai (1980) report on "hypothalamic endocrine angio-architechtonics" (pp. 306-311), a field that describes the computerlike, modular structure of parts of the hypothalamus and hypophysis, which mediate between external and internal milieus by means of both humoral and neural circuitry. A very large number of hormones and neuroactive substances, including a number that are affected during "shock" or stress are controlled by this system of neurohypophysis and hypothalamus (Makara et al., 1980). There is no question, then, that if we are eventually to better understand the relationship between sleep and dreams—and include in this an understanding of the relationship between stress, nightmares, and information processing (as written about so cogently by Max Stern)—we need to fathom this particular neurology as well, so that along with our precious psychoanalytic perspectives we can eventually truly understand how fear bothers sleep. But at the very least we know that the brain is composed of multiple learning systems that become alternately connected and disconnected from each other for purposes that appear to relate to REM (dream) periods of "reviewing" and NREM periods of "downloading" critical insights. Eventually, if our anxiety can be quelled, we then "recreate symbolically from a knowledge base novel reconstructions of our world" (Foulkes, 1983, p. 405) with obvious adaptive value. If our analysis of our situation tells us that our brain is headed toward a dangerous situation and we need some help, then we experience nightmares; and if our development has included "biotrauma," such that we expect no help will be forthcoming, then we may pass over into pavor nocturnus. Often critical in individual cases is whether the human environment will allow one to tap one's maximum potential for emotional growth.

We have now returned full circle to Stern and his theorizing about repetition and trauma. I agree with his final conclusion, that whereas for the individual there is no requirement of traumatic frustration for development to occur, for the species "the inevitable experience of trauma was absolutely necessary to the development of an innate program capable of generating anticipation." That is, over the course of human evolution trauma has led to physiological defenses against shock. The purpose of these defenses is to provide, first, a signal mechanism "in the form of agitated behavior (such as one has during a nightmare or pavor nocturnus attack) . . . and . . . primary depression . . ." and second, "external assistance in the face of disorganizing states of tension." This is the teleonomic principle that Stern has unraveled, and it is an example of the kind of creative interdisciplinary theorizing that I personally believe is not only valuable in individual clinical psychoanalytic work, but central as well to the continuing health and growth of psychoanalysis itself.

7

Psychoanalysis, Nonverbal Communication, and the Integration of Touch (Contiguity), Vision (Similarity), and Hearing (Sonority)

PRÉCIS

It is easy to fall into the trap of assuming that language means verbalization. We know that there are languages that do not rely on verbalization at all, such as the sign language of the deaf, but we tend to disavow the significance of their existence and also of the many modes of communicating that involve gesture, facial expression, posture, dress, and so forth. Bellugi's research at the Salk Institute has shown that language is exceedingly complex and that it is most certainly more than verbalization. Verbalization is important for the system of syntactical language, but this is merely one language mode among many.

What then is language if it is not strictly words or signs? The honest answer is that we really do not know yet. However, many scientists within a number of disciplines have been working on the problem. In a sense, language represents a cluster or family of related communicative capabilities. Chapter 7 explores, classifies, and illustrates some of these possibilities. There are several points to be made: (1) man's communicative systems seem roughly divided between those that are verbal and those that are nonverbal; (2) the nonverbal communicative modes seem to be inborn, function from birth or shortly thereafter, and serve as a language foundation on which the syntactical system is later superimposed; (3) exposure to syntactical language decisively reorganizes the brain in the direction of allowing for abstracting ability (which I believes is the basis for advanced psychological development, syntactical language fluency, and various cultural achievements); and (4) the neurophysiological basis for man's capacity for what Daniel Stern calls "a modal perception" (and I call crossed modal integration) remains to be discovered (although various speculations are made in this chapter and elsewhere in this book as to what this sensory integration process may consist of). Many psychopathological conditions probably result from disturbances in the sensory integration system.

Research in psychoanalysis, psychology, neuropsychiatry, and anthropology shares an interest in language. Each of us is born into a specific cultural milieu within which we are exposed to, and acquire skill in the use of, a native language. But no matter how proficient we become in using language, experience teaches us that accurate communication can at times be extremely difficult. One reason for this difficulty appears to be the multiplicity of communicative modes employed, which complement or supplement formal language. *Communicative mode* refers to all of the verbal, paraverbal or gestural, and expressive human behavior that is meant to carry a meaningful message. These modes are built into a total communicative system that taps much of the brain's receptive, processing, and expressive capacity.

One conclusion of this chapter is that the success of psychoanalysis hinges on the sensitivity of the participants to all nuances of communication, especially what is not verbal. Since Isay (1977) and others have thoughtfully reviewed verbal communication in psychoanalysis, this chapter concentrates on nonverbal communicative modes[1] classifies them, illustrates their role in psychoanalytic treatment, and considers some of the theoretical implications involved in such a language system. A second conclusion reached is that nonverbal communication is in some ways more fundamental (important) than verbal communication. It may well be the foundation on which subsequent verbal language is later constructed. This becomes clear especially in the review, at the end of this chapter, of linguistic contributions to psychoanalysis.

A third conclusion is that a major synthesis is possible for the work in psychoanalysis, linguistics, and anthropology regarding fundamental principles. These include the following: Freud's conception of condensation and displacement (two explanatory principles facilitating the understanding of all dreams and the primary-process activity of mind), Saussure

[1]The major nonverbal communication mode of concern to psychoanalysts, of course, is transference. I have not, however, included this subject in the present chapter because the subject deserves its own presentation. Thus, Chapter 8 considers the transference phenomenon, its mechanisms and possible origin, while Chapter 9 deals with the management of transference through the discussion of specific case material. The material in Chapter 9 should allow the reader a better chance to be convinced of the significance of nonverbal communication.

and Jakobson's conception of "similarity" and "contiguity"(two explanatory principles facilitating the understanding of all language), and J. G. Fraser's conception of "contact/contagion" and "similarity" (two explanatory principles facilitating the understanding of all myth and ritual). These three sets of two principles are really one set, that is one dichotomy (Muller, 1989; see also chapter 11). Moreover, if to this dichotomy is added insight from the work of Hermann and of Fónagy (see the next section) on the importance of sonority (intonation), then it is possible to see that each of the scholars mentioned above was probably appreciating (within his chosen domain) the contribution to mental activity of a fundamental aspect of brain organization, namely, the integration of information about "contiguity" (touch), "similarity" (vision), and sonority (hearing). This intersensory integration appears to be a decisive part of the system for man's abstracting ability and therefore the basis for language, culture, and advanced psychological development, each of which requires an abstracting ability.

THE MODES OF COMMUNICATION

Darwin (1872) studied emotions extensively, making detailed observations about their nonverbal expression in man and animals. Langer (1942, 1967), Basch (1976a, b 1983), Gedo (1984a, 1986, 1989a [including footnote 2, p. 181]), Tompkins (1962a, b), and others have further refined the role of man's inborn capacity to express and decode feelings by means of a limited number of standard gestures and facial expressions (that is, nonverbal communication) that establish fundamental sensorimotor schema. These gestural categories appear to be cross-cultural. Sacks (1985) has elegantly expressed the fascinating tragicomic consequences of the loss of the receptive aspect of this ability (prosopagnosia) in *The Man Who Mistook His Wife for a Hat.*

The major psychiatric study of nonverbal communication appears to be by Ruesch and Kees (1964), who divide nonverbal forms into three distinctive categories: sign language (which varies from gestures replacing simple words to the formal gesture languages of the deaf), action language (embracing all movements not exclusively used as signals), and object language (involving all expressive displays, including such activities as art and fashion). Their theoretical contribution, aside from taxonomy, relates to the consideration that the communicative world might be parsed into "analogic" versus "digital" forms (p. 189).

Common sense tells us that *visage* relates to a wealth of subtle feeling states, and from ancient times masks have allowed the human species to stereotype facial expressiveness artistically, religiously, and magically within a given culture. Today an actor is routinely typecast according to the particular ability of his face to portray chosen qualities. For example, comic or sinister qualities seem often portrayed in the West by choosing actors with faces that have abnormal proportions, asymmetries, sharp or angular features, or defective skin.

Speigel and Machotka (1974) carefully analyzed how movement and posture can be decoded into meaningful categories of intent, affect, and motivation. Freud's (1914c) case study "The Moses of Michelangelo" serves as the paradigm for applied psychoanalysis: in this case an artwork is conceptualized as a frozen moment in time, which together with prior and future movements forms part of a (probable) motivational chain.

Feldman (1959) approached communication from the perspective of repetitive verbal idiosyncrasies, demonstrating how these patterns reflect unconscious feelings and thoughts. Of course, Freud's (1901) "The Psychopathology of Everyday Life" represents the obvious forerunner of such insightfulness and the beginning of the systematic psychoanalysis of symptomatic acts. The aforementioned essay, along with "The Interpretation of Dreams" (1900), provides a Rosetta stone illuminating the communicative meaning of the unconscious as seen in slips and in dreams. Dreams have long been considered as communications and are frequently discussed as such in most societies.

Freud deeply believed that language represents a deep structure of the brain. His study *On Aphasia* (1891b) deals with both his most basic and most enduring psychoanalytic assumptions while at the same time examining the subject of disordered communication. Although not a linguist, Freud introduces a novel linguistic framework for understanding aphasia, namely, that of general systems theory, a fact that is rarely appreciated outside of psychoanalysis,. Following on and greatly extending this foundation has been the research of Saussure, Jakobson, Lacan, Rosen, Derrida, Schafer, Edelson, and Fónagy, to name but a few. Some of this work is explored later in this chapter. For example, Fónagy has clarified that intonation of speech plays a major role in communication of affect both in infancy and adulthood. The psychoanalyst Hermann also felt that intonation and sonority "are closely linked to the unconscious and to affect" (Muller, 1989). Thus, significantly before we learn to recognize the words of our native language, we are able to effectively encode and decode emotional communication on the basis of intonation alone, and we apparently continue to use certain sounds (phonemes) to carry particular affective meaning(s) independently of the denotational meaning of the words themselves (Steiner, 1987).

But so far we have only scratched the surface of the subject of communicative modes. Hand gestures also need consideration. Signing is used not only by the hearing impaired, Hawaiian dancers, and American Indians; most communication by hearing people as well is accompanied by stylized and/or idiosyncratic manual gestures that carry their own meaning, which may or may not coincide with what is verbalized. Such gestures are learned but not taught. Of course, these may also be standardized, as is the gesture for "you are crazy" (which differs from culture to culture but often is represented by some eccentric circular hand movement). Sometimes these gestures are hard to understand unless they are explained. For example, in Japan the outstretched hand (palm down) moved downward in quick, short steps means "come here," while a burned finger may be touched to the earlobe with the explanation that this is the coldest spot of the body).

A particularly intriguing observation that has been reported by Sacks(1989) describes how those people who are fluent in the sign language of the deaf can, after mere hours or days of exposure, learn to effectively communicate in a foreign sign language (which is presumably as different from their native sign language as Chinese is from Italian)! There are thus ways in which gestural languages appear to violate some of the beliefs that we have become accustomed to, such as the expectation that rapid language acquisition, for most people, involves a relatively brief window in time during the latency-age years.

In addition to the use of the hand, mime incorporates a more general form of gestural communication that combines imitation and wordless language. Speigel and Machotka (1974) have considered mime "a sort of basic code—a lingua franca—reaching across cultures and historical periods" (p. 43). Mime was a critical element in the metaphysics of Plato, who believed that the world (and all that is in it) represents a copy of ideal forms that are beyond direct scrutiny but are capable of being portrayed, as, for instance, by mime (p. 44). Most critical from a contemporary psychoanalytic perspective, mime seems to express or be close to unconscious processing since it involves similar mechanisms: condensation, dissociation, distortion, displacement, representation, and symbolization (p. 58).

Such gestural communication employs brain subsystems that appear to mature significantly before those employed in the use of vocalization. For example, deaf infants of 3 or 4 months of age are able to comprehend and signal back such manual gestures as the formal deaf sign for *milk* (Schlessinger, 1976, personal communication). Piaget has described the infant's readiness for the establishment of sensorimotor schema, Basch (in Leider, 1984) has elaborated on the vital significance of this for communicating and creating affective relatedness, and Mahler (Lilleskov, 1977) has stated that nonverbal communication is therefore best studied in children during the first 18 months of life. That is, in addition to exposure to formal verbal language, the earliest period of human life involves the learning of imitative nonverbal communication based on tapping at least two different kinds of distance receptors: auditory reprocessing of such things as intonation (as studied in detail by Fónagy) and visual processing, which helps children become the prototypical mimes.

Mathematics constitutes a language of scientific statement. This has been formalized in a variety of ways, including Boolean or symbolic logic, mathematical formulas, and software languages for the programming of computers. Communication has even been described mathematically within Shannon's communication theory. In fact, this is the origin of the term *bit* of information. Of such things most people are aware, but there appears to be less appreciation of the fact that at a clinical level numbers and mathematics can play a direct role in communication in some surprising ways. For example, there are reports of two idiot savant brothers who communicated emotionally with each other particularly by means of uttering numbers that proved to be larger and larger prime numbers (Obler and Fein, 1988). In addition, some languages have been used for complex, number-based messages. This is somewhat easier for Hebrew, because Hebrew letters also stand for numbers. *Gematria* is the "general term for a variety of traditional coding practices used to establish correspondences between words or series of words based on the numerical equivalences of the sums of their letters or on the interchange of letters according to a set system" (Rothenberg and Lenowitz, 1989, p. 205). Moreover, in addition to gematria per se there are visual or place-related methods of (mathematical) expression that include such things as anagrams, *notarikon,* and *temura.* In notarikon several words may be the clue for a new word; for example, the first letters of the expression "garden of delight" also spell the word "god." In temura there can be any systematic replacement of one letter for another; for example, the first half of the alphabet may be laid on top of and code for the second half and vice versa. The Jewish mystical writers, within the framework of Cabalistic knowledge, experimented extensively with such communication in their Hebrew prose and poetry; the history of these efforts apparently goes back to ancient Greek and Babylonian practice (Rothenberg and Lenowitz, 1989).

ORGANIZATION OF COMMUNICATIVE MODES

At this point it will help to classify the general modes of communication. A complete list of all possible modes is neither intended nor possible at the present time. Rather, the table of modes[2] that follows is a practical and

[2] The basic scheme for classifying modes of communication is clearly more refined than the usual psychoanalytic categories of primary versus secondary process. The outline I have followed in this innovation is that presented by Benson (1986) for the analysis of research data on aphasia. Benson contends that this division into language (or communicative) modes helps one understand an otherwise confusing body of research on language lateralization. Simply stated, the brain lateralization characteristics for each of the communicative modes is differ-

TABLE 1
Organization of Communicative Modes

1. Gestural mode (including "signs" as defined within a given culture or within a given language of the hearing impaired)
2. Prosodic mode (included rhythm, inflection, timbre, melody, and a system for expressive facial gestures)
3. Semantic mode (includes verbal meanings, concepts, and visual meanings)
4. Syntactical mode (includes sequencing, relationships, and grammar)
5. Gematria (includes gematria per se, as well as anagrams, notarikon, and temura)
6. Other modes (includes modes not explicitly mentioned but creatively employed by individuals or groups)

representative guide. Later we will consider briefly some of the theoretical implications behind the chosen classificatory scheme.

Before illustrating the modes with clinical examples and discussing the theoretical implications of the proposed scheme, it may help to make two additional points about communication. First, claims have been made that humans can use a "sixth sense" or extrasensory perception to obtain knowledge about feelings and thoughts. It seems possible that such an impression that there is more to human communication than formal language exchange and sensory perceptiveness can be explained in terms of the operation of the total communicative system described in this chapter. In other words, perhaps the postulation of an extrasensory perception is really an acknowledgment that our brain is continuously processing multiple modes of communication. Simultaneous with our formulation of semantic and syntactical constructions are various kinds of gestural and nonverbal communication involving the face, hands, and the entire body. Even complex mathematical, anagrammatic, and dramatic (e. g., transferential) subtleties may ride piggyback on the message transduced. The ultimate message would clearly be a composite, and it is reasonable that some substantial amount of this information coding and decoding is occurring totally outside of our conscious awareness and control.

The second point is that our preferred modes of communication and their idiosyncratic qualities or uniqueness is an important aspect of our personality or character. Such patterning would also include different interests in data communication and collection so that over a lifetime

---

ent. From our perspective, it should be apparent that the brain supports a system of multiple simultaneous language (or knowledge) subsystems that somehow become unified or integrated with each other, possibly by way of prefrontal vertical mechanisms (see chapter 4).

knowledge base reflecting one's communicative bent will be accumulated. A fuller discussion of the implications of these comments, however, is beyond the scope of any concise presentation of ideas on communicative modes.

VIGNETTES OF VARIOUS COMMUNICATIVE MODES

The necessary brevity of this clinical section will not permit a sense of conviction but should properly illustrate some of the many communicative possibilites. A fuller discussion is provided in chapter 9.

Case 1. A shooting gesture. A young lady in analysis was talking in a detached manner about recent events while her left hand was making the gesture of a gun, quite out of her awareness. The analyst's response was to call this gesture to her attention, which led to her recognizing her own anger over the anayst's plan to take a vacation that would interrupt the analysis. She had defended against this anger by dissociating herself from the treatment, and the affect that she specifically disavowed was the complex of feelings that she had about "being abandoned" and "unprotected." She then remembered how in her a childhood her parents had gone away and she had been left with a caretaker who had sexually molested her—a memory she had deeply repressed until this moment in the treatment.

Discussion. In this example it is obvious that the analyst's role was, first of all, to correctly perceive the gestural message about anger that occurred outside of the patient's awareness. I presented this case in chapter 2 as an example of the defenses of disavowal and repression defined in novel neurophysiological terms as interhemispheric communication blocks in different directions.

Herman Serota (Lilleskov, 1977, p. 704) has astutely reminded us of how difficult it is to know when to convert nonverbal to verbal dialogue in order to achieve insight. He is of course correct, and only one's entire training as an analyst can facilitate such decision making. However, it seems that one moment to consider commenting on such nonverbal communication might be when the analyst feels strongly that unless such a comment is made, the patient will lose track of a critical affective–cognitive correlation (usually as part of a major transference). Such was the case in the example presented here. Without the intervention, it is likely the patient would never have become aware of her angry affect or of its correlation with the analyst's plans for a vacation.

Case 2. Intonation of voice. A patient was observed to shift intonation very subtly during a rendition of a story that she had told several times before in the analysis. This time seemed different, however. She was asked about the unusual intonation and quite unexpectedly told the analyst to "go fuck yourself!" This turned out to be the beginning of the patient's and the analyst's shared awareness of a muliple personality structure within the patient, one for which the critical differentiation between personalities was often a shift in tonal quality. The analyst learned with this patient the importance at times of closing his eyes so as to be able to properly focus on subtle shifts in intonation, which carried the critical affective message. His awareness of this changing "musical" dimension of the patient's vocalizations was important to the analysis.

Discussion.. Furer has pointed out (Lilleskov, 1977, p. 700) the need to be careful not to assume that nonverbal communication is more primitive because speech develops later. In this example the patient's different personalities represented conflicts from various age levels of psychosexual development, although they each appeared in connection with unique tonal shifts or other nonverbal idiosyncratic messages. Some will conclude that because the patient suffered from multiple personality, this example is proof that such nonverbal communication especially goes along with more serious and earlier psychopathology. I think that although it is easy to make such a conclusion, it is not necessarily correct: nonverbal modalities, as Furer has argued, can represent any level of thought and language.

Case 3. Enactment using total body positioning. A patient who would, under ordinary circumstances, move about slightly during the analysis entered a phase in which he stopped moving entirely while on the couch. At times he would add to this either silent periods or sudden exaggerated tonal shifts in which he would start the first word of a sentence quietly and then become loud during the second or third syllable (as though he had been surprised by something). When asked about this pattern he was at first unaware of any possible significance, but as this phase of the analysis proceeded, it became apparent to both of us that we were moving toward his revelation of a critical childhood fantasy that involved his being a prisoner of the Nazis and placed in a building that was set on fire. No one would escape: movement was impossible. The inner "darkness" became terrifying. The analysis was a complex one, but part of his history included the fact that some of his family had indeed died in exactly this manner during World War II, facts he had learned about in childhood and that played a major role in his own development. The analyst's curiosity about the details noted in the analysand's shifting vocalizations and movements (or paralysis) made a decisive positive contribution to the progress of the analysis.

Discussion. Using microanalysis of videotapes of mothers and children taken simultaneously, Trevarthen (1979, 1985) has described the pas de deux of mother and children. Furer (Lilleskov, 1977, p. 703) reminds us that this "choreographic synchrony" occurs to a surprising degree during any psychoanalysis. In chapters 3 and 11 I suggest that the basis for such empathic bonding lies in the simultaneous activity of multiple brain learning subsystems, the leading part of which seems to be in this case the vestibulocerebellar system. In terms of the present example, it may help to remind ourselves that one hallmark of authenticity of recall might be the simultaneous involvement of multiple sensory and multiple communicative modes that all point toward the same experimental meaning(s). In Case 3 the patient's verbal, visual, tactile, and other perceptual capacities all seem to have captured the affective horror of being the victim of Nazi atrocity.

Case 4. A whistling interpretation. Gedo (1978, 1984b) has presented an analysis in which one of the important interpretations was made by quietly whistling to the patient a particular classical piece while the patient was talking. Slowly the patient became aware, first that the analyst was whistling, and then of the specific music. The music was a piece that the patient had once told the analyst her mother had practiced during a critical phase of the patient's childhood. The patient had listened, playing at her mother's feet, and now the old melody told her, as no other communication might have done, that what she was feeling was a repetition of this early relationship with her mother, which had been anchored in her musical memory.

Discussion. In chapter 1 I attempted to explicate the effects of interpretations such as Gedo's in terms of some possible neurophysiological mechanisms that involve intersensory bridging within certain systems of the brain associated with long-term memory storage. For the purposes of the present discussion I wish merely to note the critical importance to the patient of her analyst's musical sensitivity. He communicated to her in her own language, in a deeply unconscious and personal sense. Conversing in people's own language means much more than merely employing their native tongue. It also means appreciating the ways in which they use their language and the affective experiences that, for example, enliven their interest, create boredom, or make them angry or unglued. One thing that happens when we focus on good communication is that our brain seems to become less involved in an artificial or intellectual manner and therefore more free to experience the patient's current affective state. Somewhat differently, proper understanding on our part facilitates a proper attunement to the patient (whether we choose to respond any further or not).

Case 5. Silence can be loud or soft. A patient's silence was associated with a rigid body posture, tense breathing, and markedly inhibited movements. Previous experience suggested the probability that she was experiencing anger, which she was able to confirm when asked specifically about this. Despite a number of suggestions, it was not possible to determine the source of this anger until the following session. Then the patient could report that she had greatly appreciated my attempt to reach her emotionally across the chasm created by her angry inhibition. She had had a number of thoughts she wanted to mention but had been embarrassed to discuss them and had ended up feeling that she "never gets anything right." Now she was able to discuss them, however, because my breaking the silence had diminished this self-criticism markedly and had aroused her own curiosity about the impasse. After she shared what had been hidden, we were able to begin to speculate about a possible explanation for the hesitation. She remembered often having to wait in childhood until her mother first took care of some business of her own. Her job, she had felt, was to determine her mother's needs and see that they were satisfied first. Her own needs became postponed indefinitely, and in this sense she had the feeling of having prematurely given up her childhood. Now, in analysis, it was startling for her to appreciate that the time was really her own and that someone else would make the effort to help *her* with her own problems.

Discussion. It is all too easy to misread silence and return it with silence without carefully checking to see if one's assumption about the meaning of the silence to the patient is correct. In my experience this is an excellent example of behavior that is nonverbal yet critical communication. The analyst is not free to establish standard responses to such behavior without paying, at times, great penalties, which of course ultimately harm the patient.

Case 6. Mixed messages. A very intelligent patient always felt unable to understand even basic communication. This was traced to a tendency on the part of her father (her only parent) to communicate to her with profoundly mixed messages. When, as a child, she would reach out to him emotionally by describing her rich inner life, he would tell her repeatedly: "Don't feel sorry for yourself; you are taking yourself too seriously!" His tone of voice suggested caring, but the timing of the messages, which were profoundly disruptive to her, told her something different. She felt he was being sadistic and became so introverted that by the time she began treatment she had very little hope of trusting others with her thoughts and feelings. The intervention that helped her the most in treatment was for the analyst to carefully monitor his own messages and, when he found himself in a mixed communication (which she invariably discovered), admit his mix of feelings.

Discussion. One of the crucial reasons for appreciating nonverbal communication is that this mode can be used to express different and even contradictory messages on the same communicative channel as the verbal message. This can be profoundly confusing for analysand and analyst alike and can lead to stalemates unless recognized and appropriately addressed.

DISCUSSION OF THE CLASSIFICATION SCHEMA FOR COMMUNICATION

The communication system of the brain has been considered in this chapter to be a product of several communicative modes. That these represent different subsystems of the brain is clear from the different lateralization characteristics of these different modes, as pointed out by Benson (1986). The gestural mode is a clearly bilateral system whereas the distribution of the prosodic mode function is more complex: timbre is bilateral, melody is essentially right-sided, and rhythm is primarily left-sided. Regarding the semantic language mode, verbal meaning is left-sided, visual meaning is right-sided, and concept formation seems to be bilateral. The syntactical system is interesting in that it is entirely left-sided; that is, the right hemisphere appears to have virtually no ability regarding sequencing, relationship appreciation, or grammar. Benson does not consider mathematical modes in his schema, but gematria is probably a bilateral function, depending upon which mathematical skills are involved; in the case of highly logical functions the left hemisphere is probably most critical whereas anagrams would presumably involve visuospatial functions and therefore the right hemisphere (in right-handed subjects).

It goes without saying that a substantial portion of the brain is devoted to the decoding, processing, and encoding of communication. The brain responds to this experience by changing its organization in a process called learning, which involves a brain property called plasticity. And the permanent learning and plastic changes that occur over time are considered psychological development. *Clearly, communication, learning, plasticity, and development all represent abstractions about related psychophysiological processes.* I have been studying these processes and have made tentative correlations between psychoanalytic and biological perspectives on learning (see chapters 1 and 2), brain plasticity(see chapter 3) and development (see chapter 10). I have written (see chapters 4 and 5) about the role of the prefrontal cortex regarding language and related activities (e.g., judgment, selective attention, and adaptive decision making). The present chapter is my attempt to round out this effort by delving further into language by means of a consideration of the phenomenon of nonverbal behavior.

I hope that, by this point the reader will agree (from the explanations and illustrations given previously) that there is a reason to consider that nonverbal communication is clinically relevant to psychoanalysis. I would now like to indicate the importance of this subject to both psychoanalytic theory and neuroscientific research.

Let us start with the observation, mentioned in the introductory remarks, that those (hearing or deaf) people who know the sign language of the deaf fluently are able to learn the native sign language of those deaf people from another country in an amazingly short time. This fact suggests that the class of nonverbal communication called sign language is closer to the basic linguistic code (or so-called Chomskian "deep structure") than are formal verbal native languages. This is an important clue about a fundamental process that underlies all of the brain's abstract activities. The reasons will become clearer after a bit more discussion.

If we also consider the evidence from biology regarding the evolution of language, as explicated by Lieberman (1984), there seems to be no question that man falls in an evolutionary sequence in which nonverbal language precedes verbal language as a major mode of communication. All animals have organ systems that are "tuned" for the encoding, transmission, reception, and decoding of species-specific communicative messages. Crickets communicate with other crickets by rubbing their legs, and bees communicate with other bees (about such things as the location of and distance to a pollen source) by dancing (Moffett, 1990). Most probably, early hominids used verbal intonations, facial expressions, and gesturing before their voice box was sufficiently adapted to employ spoken language as we know it today (see chapter 2). Most importantly, from this perspective, verbal communication is not superior to nonverbal communication any more than English is superior to the sign language of the deaf.

Because we are left-brained and right-handed, the verbal aspect of our mental activity often gets a dominant share of recognition, as though logical, language-related output is more important than other (meaning, nonverbal) expressions of inner experience. Any serious study of culture will show us immediately that nonverbal expressions of affect, intent, and inner experience are equally profound and have been recognized throughout history. Who would claim that Michelangelo's Pietà in St. Peter's is less expressive than Shakespeare's *Macbeth?* Who would say that Schumann's Piano Concerto in A is less informative than *The Purloined Letter?*

And yet, when one reads about psychoanalytic interpretations in our scientific literature, it is almost invariably assumed that verbal communication and verbal interaction are all that is important. Theoretically, it would be difficult to prove that a pure verbal communication is even possible, yet nonverbal communication seems to have been systematically excluded from study. This chapter reminds us that many modes of expression combine, and in complex ways; most crucially, they can at times carry different messages (as in Case 6). The intonational, facial expressive, and signing gestures eventually evolved into human speech and languages; but the voice, face, hand, and entire body have continued to provide a richness to our words that help us overcome ambiguity and may even facilitate the determination of authenticity or fraud.

A second major part of language is that its acquisition seems to affect psychological development in a decisive way (see chapter 4). If we are to understand mind and brain then we must understand how this comes about. That is, our psychoanalytic theories should also take into account development of such major ego functions. It should ask such questions as, Why is language so critical for psychological development? Why is it that, whether we are deaf or hearing, if we have a timely input of native language (English or American Sign Language of the Deaf), then we end up being able to experience a normal cognitive developmental pattern, but if we are deprived of this input, we can very easily end up functionally retarded?

Although it is not yet possible to answer this question, I think it is worth making an effort to do so. In my opinion the answer lies in appreciating how "abstractability" becomes possible as a mental function. The key to our learning how to abstract, (i.e., as a permanent ability) lies in training the brain to make connections between a number of processes. On one hand, the verbal and the nonverbal must be brought together.[3] This happens only through exposure to the naming experience, which is how parents share feelings with their offspring (see also Muller, 1989). Words are given to experience; for instance, "the ability to evoke the immediate experience of the spatial relationship 'above/below' by the words 'sky/earth' in the absence of an actual view of the sky and earth hardly seems like advanced reasoning capability. It is however . . . [the critical] step in abstract symbolic logic. . . [because] we gradually acquire words for ever more general and abstract categories, and ultimately give names to actions and even concepts" (Kent, 1981, p. 149).

The fact that we have names for complex patterns of thought, as well as for experience that is perceptually near, gives us the ability to begin to manipulate, communicate, and disseminate meanings. But an additional integration must occur that appears crucial for both language

[3] Of course, another sense in which verbal and nonverbal are brought together is discussed by Freud (1915) in his essay "The Unconscious." where he describes connecting systems of unconscious and preconscious. The reader will notice, however, that Freud focuses on the unlocking of repressed memories in adults whereas the current discussion concentrates on verbal-nonverbal conscious integration as an early developmental process in infants and children.

and psychological development (and possibly all abstract brain activity), and this is the integration of the various sensory modalities.[4]

The major learning-related, knowledge-acquiring subsystems of the brain (see chapters 1, 2, and 3) hinge on the process of crossed sensory integration. [5] The linguists have carefully analyzed the part of the experiental universe of interest to them and have concluded that the abstracting activity we call language is basically the combination of two fundamental principles: "contiguity" and "similarity" (Jakobson, 1987). This generalization represents the great achievement of Saussure and Jakobson, later carried by Rosen (see end of this chapter) into psychoanalysis. Coincidentally, it also represents an insight of Freud (1900), who covered the two basic primary-process mechanisms of condensation and displacement

[4] This discussion as well as what follows is intended to describe neuropsychologically only one, albeit an important, aspect of language. Clearly, however, the basic mechanisms of language are extraordinarily complex, and I do not want to convey the impression that we are anywhere near understanding how the brain accomplishes the coding, processing, and decoding of language (or communication in general). Those interested in scholarly analyses of some of the central problems in this area might wish to consult Goldman (1986), Boden (1988), and Shallice (1988).

[5]By intersensory integration I am not referring to the same phenomenon that Stern (1985) describes as "amodal perception" (nor synesthesia, also described by Stern). Amodal perception is our ability (appearing shortly after birth) to take information received in one sensory modality and somehow translate it into a different sensory modality (p. 51). Synesthesia is the ability to experience one sensory modality in terms that are associated with a different modality (e.g., hearing music in the form of shifting "colors" or "colorations"). In contrast, I am referring to the integration that occurs in those areas of tertiary cortex that receive sensory input from the primary association cortical areas for the major sensory modalities (touch, vision, hearing, etc.). The central parietal cortex is only one representative area within the brain that accomplishes such an integration process. It should be apparent, however, that Stern's amodal perception concept already represents a stage in which some abstracting has already occurred, in the sense that the original sensory information is assumed to be broken down into abstractions about time, form, shape, etc. (presumably as a prestage to being translated into another sensory modality format). For Stern this ability for amodal experience is vital because it becomes the basis for the infant's unified experience of others and the "emergent sense of self" (p. 52). The present chapter on nonverbal communication, however, does not wrestle with the issue of self formation but rather with more basic questions: What is the process of the integration of sensory modalities (within such areas as the central parietal cortex), and what are the implications of this blending process for psychoanalysis, especially as regards the emerging ability to abstract and the subsequent developments (psychological, cultural, linguistic) that are dependent on this ability?

(which are analogous to contiguity and similarity,[6] and of Fraser (1890), who in his monumental study of ritual and myth came to the conclusion that the two critical organizing principles of mankind are "the law of similarity" and "the law of contact (or contagion)" (Muller, 1989, p. 102). What is most fascinating is that *it is possible to combine these psychoanalytic, linguistic, and anthropological insights about the same two "fundamental principles[7] with the work of Hermann (1934, 1936) and Fónagy (1963) on the importance of sonority,and see that the great dichotomy itself is most likely based on the brain's integration of information along three axes: contiguity (touch), similarity (vision), and sonority (hearing).* Those readers familiar with the mapping of the central parietal cortex will appreciate especially the beauty of this, since this parietal sensory integration zone is placed equidistant from the primary cortical sensory areas for the perception of touch, vision, and hearing. And it is within this parietal integration area that the perceptual equalities of experience become "abstracted" into memories that are no longer coded according to any sensory tag. In fact, *this lack of attachment to immediate sensory experience is the essence of abstraction.*

Perhaps what "naming" (and thus the timely exposure to language) accomplishes in facilitating development of both language acquisition and abstracting ability in general is the creation of connections between the primary sensory modalities based on their simultaneous stimulation (see chapter 1). This exposure would be expected to form bridges or correspondences between experience in different sensory modes based on their sharing a common "name."[8]

---

[6] My thinking about the equivalence of Freud's concepts of condensation and displacement and Jakobson's (and Saussure's) "contiguity" and "similarity" principles is also supported by Lacan (see Muller and Richardson, 1982).

[7] I am aware that the two principles noted are usually considered (from the psychological perspective) to relate primarily to either visual experience or to sensory experience in general, without specification of the involved mode (i.e., that my own connection of "contiguity" with touch sensation and "similarity" with vision may appear somewhat idiosyncratic). What is intended by the chosen connections, however, is to emphasize what I feel are valid correspondences for that penultimate period before language development per se in which the child's grouping (categorization) of objects appears to be primarily decided either by scrutiny of their "visual similarity" or examination of their "tactile proximity" to each other. Obviously, the problem is complicated by such phenomena as "amodal" perception (see Footnote 4). What we need most is accurate information about the mechanisms by which intersensory integration occurs.

[8]What happens during "naming" is an exceedingly complex subject, and one additional ramification therefore needs to be noted. The thalamocortical-activating system is another potential anatomical substrate for some aspects of

THE LINGUISTIC CONTRIBUTIONS TO PSYCHOANALYSIS

To round out the previous discussion and better identify relevant clinical and theoretical issues, I will comment briefly on linguistic approaches to psychoanalysis. Since Leavy (1983) has comprehensively reviewed this subject, I will organize my remarks around his analysis of the contributions of Victor Rosen, Marshall Edelson, Roy Schafer, and Jacques Lacan, Although I am generally in agreement with Leavy's scholarly reflections and with his contention that such research has greatly assisted psychoanalysis, there are some areas of significant disagreement that will become quite clear.

As Leavy states, each of the four contributors has attempted to add something distinctive to psychoanalysis: Rosen, in the area of general linguistic concepts; Edelson, regarding Chomsky's conceptualizations; Schafer, through his "action language"; and Lacan, by means of his own unique insights within the linguistic tradition of Saussure and Jakobson.

Rosen believed that interpretations in the analytic process arise as in the paradigmatic case of "finding [the] meaning of a parapraxis" (cited in Leavy, 1983, p. 38). According to Leavy, Rosen stated that "symbolic language is rooted in presymbolic sounds—or gestures—that are fixed in meaning to specific events. They form a system of signal and sign phenomena that stand for objects of the external world, and they later resonate with the superordinate secondary process" (p. 39). This is consistent with the neurophysiological and information-processing perspectives expressed earlier in this chapter. Leavy notes that Lacan used the comical neologism "lalangue" (which equals the "lalala" of a baby plus *la langue* meaning language) to express this same concept. From this perspective, language is seen as having a disposition to break down (into infantile structures, or, the lack of structure) and "is exemplified within the system of adult language, not only in parapraxis, but in the intrusion of words that mean more than they are consciously intended to mean, because they have been chosen from the level of the primary process" (p. 39).

As noted, I agree with Rosen's and Lacan's conception of the possible origin of symbolic language; however, I think it is unwarranted to conclude that language has a disposition to break down as described. Rather, it seems equally possible that there is simultaneous usage of both primary-

---

language and memory (in addition to Broca's area, Wernicke's area, the prefrontal cortex, and other areas unnamed or unknown). For those interested in a Penfield-like, experimental, surgical approach to the mediating role of the lateral thalamus in language, Ojemann (1988) should be consulted.

and secondary-process language (in terms of my previous discussion of multiple, nonverbal and verbal, communicative modes) by the brain. Verbal slips, instead of being a breakdown product, would thus represent examples of language usage in which the normal multiplicity of communicative modes becomes more or less obvious to the skilled observer; but language as communication would appear always to be divided into multiple, simultaneous, verbal and nonverbal messages. And this multiplicity of communicative channels would probably seem obvious to us if we were more gifted at recognizing and decoding nonverbal language. As the haiku poet Basho once wrote, "I didn't hear the silence of the ancient pond until the frog jumped into the water!" Put differently, what the linguistically oriented psychoanalytic scholars (with the exception of Fónagy, 1983, as noted earlier) have added, at the very least, is a method for extending Freud's insights regarding the multiplicity of simultaneous messages in the verbal realm. But the narrow linguistic approach appears essentially to ignore the nonverbal and even the nonsymbolic realm.

Continuing his discussion of Rosen's contribution to psychoanalysis, Leavy (1983) states that "what is essential to Rosen's thought . . . is that there is in all analytic listening and interpretation a dialectical moment. One process is that whereby the analyst detects the infantile sign (or signal) aspect of the spoken word, the other process [is] the interpretive act proper in which the discovered sign . . . is thematically returned to the analysand" (p. 40). Lacan, according to Leavy, makes a similar distinction between so-called full speech and empty speech.[9] It seems to me that one additional possibility is that Rosen, Lacan, Leavy, and others observe the phenomenon as they do because they perceive the psychoanalytic situation from a special perspective that believes "in the beginning was the word." That is, for them words are paramount. However, observers, whose conceptual bent is less centered on words per se might observe something different; for example, they might experience the psychoanalytic situation as something that primarily involves communication between collaborating individuals. I argue in the following paragraphs that this latter, total language perspective (or emphasis) seems to me much more helpful to the analytic endeavor than one focusing narrowly on verbalization.

Even Leavy (1983) himself introduces the idea that Rosen's ideas, and some of Lacan's as well, regarding psychoanalysis suffer the "shortcoming of the application of [too much of] Saussurean linguistics" (p. 42),

---

[9]Lacan's distinction regarding "empty speech" and "full speech" apparently also shares a kinship with Heidegger's concepts of "Rede" and "Gerede" and with similar distinctions between thought and speech, respectively, by the noted linguist Sapir (Benvenuto and Kennedy, 1986).

meaning too much emphasis on the word as a signifier. But Leavy here is referring primarily to too much emphasis on the word in contrast to "living speech as it is delivered in sentences" (p. 42). Thus, the preferred subject relentlessly remains verbal communication although the units have enlarged (and by mentioning "living speech" he presumably would not exclude nonverbal elements).

Leavy believes that Edelson's attempt to employ Chomskian linguistics is essentially a failure because Chomsky's "deep structures" really have nothing to do with the dynamic unconscious. Leavy does, however, find helpful Edelson's perspective that "much of the understanding the psychoanalyst contributes to empathy, intuition, or conscious or unconscious extralinguistic information actually derives from his own internalized linguistic (and semiological) competence of whose nature and existence he must be altogether unaware" (p. 43). My contention (see Précis, chapter 11) is that this semiological function that Rosen is referring to is the same nonverbal domain that is the subject of this chapter. Leavy concludes (erroneously, I believe) that "linguistics in a larger sense *is* the basic science of psychoanalysis because it explores the processes by which psychoanalytic exchange takes place" (pp. 45–46).

To return to my statement about the advantages of seeing psychoanalysis more as total communication (i. e., semiotics) than verbal language activity per se (close though these concepts may seem, they are vitally different), it should be obvious that the strictly linguistic approach will also tend to narrow the perspective of both participants along the lines of the analyst being the expert who tells the patient what the patient's comments really mean. In contrast to this perspective, the conception of psychoanalysis as total communication invites a broadening of scope in which the collaborative process of equal participants is emphasized. In addition, it seems that a narrow focus on words and their usage is more likely to lead to intellectualization in which the forest (of affect, embedded in both verbal and nonverbal communication) might readily be missed for the trees (of verbalization or verbiage).

But my position becomes even clearer if we proceed to Roy Schafer. Schafer has proposed his "action language" as a way of conducting psychoanalytic interpretation so as to highlight the insight that nothing in human experience just happens, but is self-created by action (and, presumably, intended inaction(s) as well). Of course, this merely accentuates Freud's original insights. Although applauding Schafer's attempts to breathe life into what he sees as a stagnant psychoanalytic theory, Leavy (1983) ends up nevertheless wondering "whether the practical proposals of Schafer amount to more than a [mere] grammatical housecleaning" (p. 48). I agree. However, as noted, Leavy's criticism of Schafer might equally apply to much of Leavy's description of the psychoanalytic contributions of the linguistic school, for this approach tends, as I believe Leavy has accurately described, to share the danger of sterile intellectualizing in the treatment setting. This would appear to be an especially robust possibility if one actually made a serious attempt to carry out Schafer's awkward-sounding treatment recommendations!

It is to Leavy's credit, however, that at this point in his discussion he introduces a critical question: "How does the analyst know enough to make [an] interpretive comment?" in the first place (p. 51). The question cannot be answered, but Lacan's efforts seem most relevant, especially through his perspective of the process as "a joint operation in which the self-analysis of the analyst—and his analysis *by* the patient—is concurrent with the analysis and self-analysis of the patient" (p. 52). It is interesting that although Lacan appears in his writings to downplay affect as well as nonverbal communication,[10] nevertheless, he, through his emphasis on intersubjectivity and collaboration (also see Benvenuto and Kennedy, 1986), appears closest to my position that communication (semiology), and not language per se, is important. We certainly need the insights from linguistics on decoding verbal behavior; but perhaps even more we need a balanced view in which equal efforts are made to expand our knowledge of nonverbal communication. In fact, as Lacan has noted, this is where Freud started with his focus on the interpretation of dreams, a mostly visual not verbal message.

[10]It may be important to note that, as might be expected from Lacan, whose use of language was both creative and idiosyncratic at times, it is difficult to criticize his written comments on nonverbal (versus verbal) communication because there is a strong possibility that he means something different from my definition in this chapter when he uses the word *nonverbal*. Lacan seems to use this word to refer to aspects of the relationship between the analyst and analysand; I am using the term to mean strictly a mode of communication or information transfer (Benvenuto and Kennedy, 1986, p. 167; also see Leavy, 1984).

8

The Transference Phenomenon

Possible Origin and Mechanisms

PRÉCIS

If one had to decide what constitutes Freud's major discoveries, without which there would be no psychoanalysis, one would mention the following: (1) the dynamic unconscious, (2) the related concepts of psychological defense and intrapsychic conflict, (3) the Oedipus complex, (4) the method for interpreting dreams and parapraxes, and (5) the phenomenon of transference. This book has considered novel aspects of each of these basic insights. Chapters 1, 4, 5, 7, and 12 consider the out-of-awareness organizational-motivational systems of the brain, which will eventually be sufficiently known so that a clear correspondence can be created with Freud's system unconscious. Chapter 2 specifically sets out to define one kind of psychological defense in terms of neurophysiological mechanisms (in this case, communication blocks between the two cerebral hemispheres). The Oedipus complex is considered in chapters 2 and 3 from the novel perspective that this period (from age three and one-half to five) also coincides with the myelinization process that ultimately bridges the two hemispheres into one overarching system. It may not be an accident that this watershed psychological period occurs at the time of major functional evolution in terms of a bihemispheric collaborative system. Dream and sleep research are discussed in detail in chapter 6 in an attempt to correlate these two areas with an information-processing theory of REM/nonREM sleep. And the present chapter, as well as chapter 9, attempts to explicate aspects of transference.

Freud's perspective on transference seems uniformly personal, that is, ontogenetic. It is in one of his recently retrieved lost papers on metapsychology that the phylogenetic perspective on transference appears,[1] so far as the author is aware. For this reason, it seems fitting to wonder about the role of the transference phenomenon in the species. Basically, the conclusion can be summarized as follows: the phenomenon of transference gave early man survival value by (1) increasing the chances of his older, unsatisfied needs being met (as when persons in the here and now respond in a "complementary" fashion to our transference enactments) and (2) providing the chance of a comparison between at least two different affect states of transference versus nontransference, thus allowing the person to observe the different possible responses of the human environment to a particular transference (that is, others might act in a way complementary to our transference or in a noncomplementary manner) and potentially to learn from this comparison. In these two ways I have referred to transference as a "probe", much as certain DNA or chemical probes exist that help the chemist identify what is in some unknown batch of genetic material.

One final word: Those readers who do not yet know from personal experience what it feels like to be undergoing a transference may find it

---

[1]The recently discovered paper of Freud's on metapsychology (Grubich-Simitis, 1987, reviewed in Spruiell, 1990) considers transference in nearly the sense intended in this chapter, that is, as a phenomenon (versus specific transferences per se). In a letter sent to Ferenczi July 28, 1915, Freud apparently included a manuscript that he called "A Phylogenetic Fantasy," which deals with his idea that the human mind evolved (transference phenomena) in stages, two of which cluster around the struggle for survival during the Ice Ages. In the first stage (with the greatest danger) man passes from reacting intrapsychically (as in anxiety hysteria) to reacting with repression (meaning adaptively or defensively; compare conversion hysteria) to reacting with obsessional organization. During the second stage, man enters into a social period with the sequential appearance of narcissistic neuroses, schizophrenia, paranoia, and "melancholia-mania" (p. 119) as remnants of the preceding history (specifically, as the consequence of mankind's collective recollection of the primal father's castration and exile of the younger males of the horde). The point in mentioning this paper, later destroyed by Freud, is to note that although he clearly did not belive sufficiently in the specific Lamarckian narrative to see it published, Freud did nevertheless consider seriously the possibility that transference, as seen through its various clinical categories, reflects or captures man's adaptive history. For example, the developmental line involves the establishment first of the ability to organize memory around anxiety, psychological defense, and cognition; then later on certain narcisisstic developments occur. Those interested in the details will wish to consult Grubich-Simitis' fascinating discovery.

difficult to appreciate my abstract arguments about the origin and mechanisms of the phenomenon. If you are such a reader, I ask that you imagine that you are lying on your analyst's couch. Assume as well that you are beginning to feel a tendency to regress in the relative safety of this setting. You are talking, following your free associations, and yet now intrusive thoughts start to appear. You momentarily believe that your analyst (who has never done anything to hurt you) seems about to attack you, or criticize you, or laugh at you, or fall asleep during your hour, or some such thing. You turn to look at the analyst, only to discover that he or she is perfectly attentive and behaving quite appropriately. This is what proves to analysands over and over again the power of the unconscious mind and invites a search for the origins of such fears within the patients themselves. This chapter is not about specific psychoanalytic transferences but rather the transference phenomenon itself: its possible mechanisms and origin. The transference phenomenon, which Freud describes so eloquently, will be seen as representing a deep structure or strategy of the brain, one with biopsychosocial adaptive value. In the following four-step argument, I will attempt to delineate transference as a phenomenon.

The argument begins with the idea that transference bears a relationship to the brain's abstracting activities, and therefore to understand transference one must appreciate how perceptions lead to conceptions; in other words, one must appreciate something about the abstracting function itself. A second step involves a thought of Einstein's about the theoretical relationship between perceptions and conceptions, a problem in philosophy and neuropsychology that has been of interest to some of the finest scientific minds.

Einstein stated that the relationship between sense experience and concept "is not analogous to that of soup to beef, but rather of a check number to overcoat" (Holton, 1979, p. 189). I interpret Einstein's comment to mean that whatever the brain does in creating abstractions from the building blocks of sensory experience, the conceptual by-products get tagged for future reference (i.e., within the brain's memory subsystems). But the tagging system itself is *not* a reflection of (that is, it has no intrinsic logical-biological relationship to) the original sensory experience (i.e., it is not related as soup is to beef). Rather, the tagging process is arbitrary and practical, for example, like the tagging of overcoats in a museum.

Einstein made another point that I wish to introduce here, one which relates to discovery in general. "Thought experiments" allowed him to combine ideas playfully, steering a course between twin dangers: the high ground of logical deduction and the low ground of empirical observation (induction). Einstein was deeply suspicious at times of both. One such *gedankenexperiment* of Einstein's was to conceptualize someone who is free-falling from a house roof (Holton, 1979, pp. 156–158). From the perspective of such a person, gravity does not exist (in his vicinity). This is because any other falling objects near him will be falling at the same speed and therefore will not appear to him to be falling at all. (The fact that at the end of the fall gravity will again assert itself does not detract from the logic of this example, which is only meant to obtain for the period of the fall itself.)

The third step of the argument involves imagining the brain as composed of modular elements whose purpose is to order experience along a time line. Time will be expressed visually by our thinking of various levels of the developmental sequence from infancy to adulthood, parallel to each other, as are the floors of a building. Much of the work in cognitive psychology, neuropsychiatry, and psychoanalysis has conceived of such a multileveled organization of mind; where the theories differ relates to how the levels are characterized and the developmental sequence understood. Figures 1 and 2 of chapter 1 represent my attempt to portray such a developmental model spatially, integrating the perspectives of Piaget, psychoanalytic theory, and the brain research of Lassen, Ingvar, and Skinhøj. This chapter continues the effort on the basis of similar assumptions, to describe the role of intersensory integration.

We are now ready for the fourth and final step. Consider the following gedankinexperiment: imagine yourself free-falling through that part of your own brain that represents the various parallel levels of your own (psychological) developmental hierarchy. Such a free-fall would stand for the transference, experienced from the perspective of the free-falling (regressing) subject. The image thus captured can now allow us to consider a number of aspects of what happens to someone during the transference experience, that is, from its own unique perspective. I hope the image will also allow us to correlate this experience with some possible neurological mechanisms.

One aspect of the free-fall is that just as gravity does not exist for the free-falling subject, so in a similar manner our own transference states will not ordinarily be apparent to us. The only perspective that systematically encourages persons to become aware of their own transference(s) is the framework of psychoanalysis.

A second major aspect is that during the clearly regressive process involved, the level we reach (in free-fall) becomes apparent by the particular models of self-in-the-world that are activated. That is, when we enter each level, the current model of self is replaced with a different model appropriate to the new level.

A third, but less obvious, aspect of the transference phenomenon is that a decisive determinant of human life would thus be the construction of such time-tagged models, the manipulation of which *is* one kind of thinking. In chapter 3 I presented a theory regarding the phenomenon of "operating in cerebellar mode" but would like now to suggest that during development the cerebellar self-in-the-world model is complemented by self-models of increasing complexity (since the brain continues to myelinate through most of the life span).

It is further postulated that during an analysis these internal models, or their operating systems, change (see Chapter 5 regarding operating system control over input/output relations, especially regarding the role of language). In a nutshell, such psychoanalytic theories as Gedo and Goldberg's (1973) hierarchical developmental model reflect in their very structure the hierarchical arrangement of the brain's own organization. What changes in the case of each phase of development is the organizational principle serving as a fulcrum for organizing.

The tendency for certain transference free-falls to diminish or to become more rapidly cycled and resolved during a successful psychoanalysis (as Schlessinger and Robbins, 1975, have described) suggests that the core changes occurring during the process involve a complex reorganization of the brain. Such "learning", however, involves so many potential changes within different levels of the brain's plastic, hierarchical structure that any analogy to computer reprogramming is a gross oversimplification indeed.

But what is the transference phenomenon, really? Why does it occur, what is its basis within the brain's design, and what is its adaptive value to the organism? Some tentative answers are possible, based on a theory introduced in Chapter 7 of this book. This theory can be stated concisely, but readers may wish to review the complex supporting evidence and detailed reasoning. The theory describes the brain's information-processing strategy that underlies REM and nonREM cycling, thereby bridging sleep and dream research. During sleep the brain is not required to pay as much attention to danger, because we are essentially immobile and less apt to get into trouble. The brain uses this opportunity to process the residue of recent experience. As Freud described, the dream part of sleep relates to wishes and what interferes with wish fulfillment. This cognition is accomplished by means of cycles of REM sleep (associated with dreaming) while nonREM periods are associated with more purely thoughtlike activity. The major unanswered question then becomes, Why does the brain need REM/nonREM cycles?

The answer provided in Chapter 7 states that REM periods are required for the three (or more) major knowledge bases of the brain (the corticolimbic system, the corticovestibulocerebellar system, and the corticostriatal system) to consolidate their input from the day's sensory stream of data. Supporting this function (during REM periods) there appears to be a relative disconnection of these same major information-processing systems from each other. Following this REM or consolidation period [2] there are nonREM periods in which these three knowledge bases appear to become reconnected. They are now in a position to "download" their data to each other. From my perspective, sleep consists of alternating periods of consolidation and downloading, which keep the major brain knowledge bases aware of the day's sensory input stream from a perspective integrating the unique viewpoints of each. The reader may ask, what advantage accrues from such cycles of data storage and transmittal between the knowledge bases? The answer is that each of these three (or more) systems is poised to extract *different* information from the same sensory stream. Therefore, the product representing the combined (integrated) input of these three systems, like the integrated product from the visual input from our two eyes (rather than the input from a single eye), allows for the creation of a master database with greater depth of field, so to speak.

Having considered transference mechanisms, we may now ask again about the origin of the transference phenomenon. Let us consider a second gedankenexperiment. What would it be like if some form or derivative of the fundamental pattern of REM and nonREM sleep cycles were to become superimposed on (or continue into) the state of wakefulness? (See Kleitman, 1963, and Klein and Armitage, 1979; also see p.69 of this book.) How would one be able to recognize such an actuality? (Note here that the expression "some form or derivative of" in the previous sentence is critical, because we know that the REM phenomenon itself does not usually continue into wakefulness, except possibly during the pathological condition called narcolepsy.) In answer to the question, it seems possible that *we would see the information processing strategy which I assume to underlie the phenomenon of REM/nonREM cycles appear during wakefulness in the form of cycles of transference and nontransference states*. From this perspective, the state of transference would coincide with a REM-like or dreamlike experience, that is, an experience in which one is not aware of the activity as it occurs and in which the usual rules for reality experience would temporarily be suspended in favor of more archaic rules or models. The nontransference state would correspond with the nonREM period. Most importantly, the pattern of REM/nonREM cycles and transference/nontransference cycles would indicate that in both wakeful states and in sleep the same fundamental brain strategy would be active as a major method for analyzing sensory data: alternate cycles of "consolidation" and

[2] There is evidence that long-term memory stability and maintenance are dependent on paradoxical (that is REM) sleep occurring over a protracted period of time (Gutwein, Shiromani, and Fishbein, 1980, quoted in Rosenzweig and Bennet, 1976, pp. 272–273).

"downloading" (involving essentially a comparison of experience from at least three different perspectives, represented by the three or more major knowledge bases of the brain). This might be compared with the idea presented in chapter 8 on the role of intersensory integration of the three primary sensory modalities (along with "naming") in the creation of abstractions. However, at the highest level the integration would not involve the combination of information from the three primary sensory modalities but rather the complex integration of corresponding data within three entire knowledge bases of the brain.

To summarize, I contend that the transference phenomenon (if considered within the pattern of cycles of REM/nonREM during sleep and transference and nontransference states during wakefulness) itself would be a reflection of an adaptive process in which successively more and more abstract "abstractions" are produced from the stream of sensory input by means of integrating and/or comparing the perspectives of the major databases or learning subsystems of the brain in a reiterative process. Long before psychoanalysis understood the significance of the transference process and therefore brought it potentially under man's control, this phenomenon of transference was part of the brain's system for learning from experience. Transferences of different types result either in complementarity and the meeting of primary needs or in noncomplementarity and the formation of new judgments about the utility of old strategies for obtaining need gratification. Any ancient hominids who were capable of such transference "probes" had brains operating with a sophisticated system for self-evaluation that would have conferred an adaptive advantage.

Before ending, it seems vital to answer some questions that discerning readers will already have thought of. If I am suggesting that transference and nontransference cycles might follow a pattern of REM-/nonREM periods, it will help to discuss more thoroughly some of the evidence supporting such a conclusion (aside from the logical arguments, based on analogies, stated earlier).

We will begin with the repeated clinical observation of every practicing psychoanalyst that dreams give concrete expression to the transference (Altman, 1975, p. 93) although this does not necessarily mean that transference states and dreamlike states relate directly to each other; their similarity might merely reflect a common origin. In this sense, it is only suggestive but not convincing evidence that there are many other structural similarities between REM (dream) states and transference states, which include the following: both represent ego regression; both make use of both primary and secondary process mentation and involve such similar mechanisms as condensation, displacement, and symbolization; both rely on nonverbal communication more than verbal communication; and both involve repetitive behavior in which there is a compromise, with simultaneous expression of wish and reactions against wish fulfillment.

The most convincing argument in favor of my theory, however, is the evidence that some dreamlike state or mode is known to extend into wakefulness and has been observed and pondered over by psychoanalytic theoreticians for many years. George Klein (1966), in an article on the "several grades of memory," reviews his own work and that of Rapaport[3] and Rubinfine and claims that specific states of consciousness or ego states accompany both the stage of memory acquisition and the separate stage of memory retrievel (p. 383). Most important for this discussion, memory recollection can occur without conscious awareness, in the form of action (p. 382).

Rubinfine (1961) sees the same phenomenon, quoting Rapaport's observation of two contrasting modes of consciousness: "monoideic consciousness" of drive gratification and "polyideic consciousness of . . . perceived external reality, internal needs, and memories of past experience" (pp. 85–86). To Rubinfine these varieties of consciousness eventually become differentiated with increasing clarity during development; that is, the younger the child, the less they are differentiated. The general rule also holds that such symbolization processes are increased by sensory deprivation, which is part of the psychoanalytic technique (pp. 74–75). Finally, Rubinfine quotes Piaget to the effect that there are altered ego states that are conducive to particular modes of recollection, and "the semiconsciousness of the dream is indeed comparable to the state of complete egocentrism characteristic of the baby's consciousness" (pp. 85–86).

Thus Klein, Rapaport, Fine, and Piaget are each seen as discussing the fate of certain "archaic schemes of ego functioning" (Klein 1966, p. 383) that can intrude into wakefulness, that may or may not be conscious, and that seem to represent a from of recollection (including recollection through action). I submit that this dreamlike archaic state, associated with an out-of-awareness, recall-equivalent action mode, sounds suspiciously like the transference phenomenon itself.

To summarize, what produces transference is the presence of three factors working in concert: (1) a repressed infantile wish, uncompleted or unresolved, stimulated by a day residue (experience); (2) a phase of dreamlike wakefulness that is a regressive opportunity (and that is associated with specific information processing and long-term memory-fixing cycles of the brain); and (3) the presence of a suitable object.

---

[3]Gill (1967) quotes Rapaport that "phenomenona analogous to dream mechanisms occur in waking thought. . . " (p. 291) (see also Giora, 1974). Noting that waking thought is not necessarily secondary-process thinking, Gill regards these as regression in the service of the ego.

9

Management of the Transference

A Clinical Case Study

PRÉCIS

Case studies can easily become either difficult to follow or altogether unconvincing when used as evidence. Therefore, in the following case material I concentrate on describing the course of a particular psychoanalysis and not on proving the argument stated in chapter 7 on nonverbal communication. In this manner the reader can better decide if my claim that nonverbal communication is vital to the practice of psychoanalysis seems supported.

After writing chapters 7 and 9, I discovered in a file of personal correspondence a letter dated 9/13/85 and a brief draft of a scientific paper that had come from a deceased friend and colleague on the subject of "facilitation of the analytic process by receptivity to the prosodic components of the analysand's speech." I had forgotten about this material, but its contents are worth quoting. My cryptamnesia seems pardonable, based on the need to forget the painful loss of my friend. Dr. David A. Brueckner's opening comments are as follows;

> *Empathy with the analysand's affective state is generally agreed to be important in psychoanalysis, but when the analysand lies on the couch the analyst's view of his facial expressions and other gestures (as carriers of affect) becomes somewhat limited. However, by contrast, the prosodic or musical component of his voice remains available; therefore, one can argue that a careful or refined receptivity on the part*

> *of the analyst to the prosodic element of the analysand's voice will be helpful in deepening the analytic process.*
>
> *The right hemisphere (in most left-handed people) contains in its insular-opercular region an area homologous with the speech area of the left hemisphere, and the function of this right-sided area is both expressive (motor) and receptive (sensory) prosody. The insula is analogous to the general cortex of submammalian species and is mature at birth. The opercular regions probably mature after birth. It is a common observation that infants and young children make sounds of variable pitch, rate, rhythm, and intensity (as affective signals) long before they utter propositional speech. Thus, prosodic verbalization precedes propositional communication.*

This is the first presentation of these marvelously precocious insights by Dr. Breuckner.

Chapter 7, on nonverbal communication, contained some brief examples that were meant to be illustrative rather than convincing. The following psychoanalytic case is presented in the hope that the reader may be more convinced that psychoanalysis benefits significantly when both participants are aware of nonverbal as well as verbal communication.

Dr. Z. was a 28-year-old, single, Jewish resident in pediatrics who presented with the complaint that she could not sustain relationships with men. When first met she was plainly dressed in a surgical scrub suit, with stethoscope dangling around her neck and various instruments crammed into her lab-coat pockets. Her intelligence and sense of humor came through immediately, along with significant anxiety (which we identified later as relating to her concern about whether or not I would find her acceptable for an analysis). During college she had sought out psychiatric help (for "moodiness" through the school infirmary, where she had a brief psychotherapy experience. However, she had experienced the therapist as overly critical (for example, when he commented that she seemed to be "sexually promiscuous"). I suggested that she might have similar expectations of me; that I would be critical or that I would not let her feel comfortable enough to experience her feelings. She remembered that he did not have tissues in his office. The fact that I did suggested to her from the very first session that it was all right to cry with me. The crying began as soon as she told me about her father.

Her parents were survivors of the Holocaust, having reached Canada in the years immediately following the Second World War. In an environment of Yiddishkeit and lingering fears of the gestapo, Dr. Z. was born and grew up in large metropolitan area. There were two sisters, five and seven years older than the patient. Before going to a concentration camp her mother had given birth to a daughter who subsequently died of starvation. Eventually the family moved to the American west coast because of the father's work in the movie industry.

Dr. Z. knew her life had not progressed in some manner since the death of her grandfather when she was 19. Her mood plummeted, and after a brave attempt to continue her studies she had had to drop out of a highly competitive academic program at a prestigious university and transfer to a smaller college of liberal arts. The "failure" did not seem to be because of intellectual or creative deficits. In the new academic program, Dr. Z. began to succeed again, finding an interest in science, but she had a tendency to get into difficulties with authority figures. She also felt inhibited sexually in the sense of not being able to combine being loving and being sexual with one and the same person. She thought she wanted to be a doctor and prepared herself for this profession, but in spite of an adolescence marked by determination and even leadership potential, she often felt "lost". She seemed to have achieved only an "illusory independence." These were by no means all of her presenting symptoms but constituted what she herself singled out.

I felt positive about her and decided we could collaborate on her analysis, an the basis of her emotional aliveness and active fantasy life, her introspective capacity, her ability to tolerate painful affects, and her lively intelligence and strong motivation for treatment. A tentative diagnosis of hysterical neurosis with some depressive and phallic-narcissistic features seemed right; that is, the assumption was made that the major fixation was at an oedipal level. At that time (age five) the patient's aunt died, and her father became ill with stomach cancer, to which he finally succumbed when the patient was seven year's old. The aunt was in many ways a mother figure, not only for the patient but for her mother as well. A detailed history supported the possibility that Dr. Z.'s experience of the oedipal situation had been one of intense guilt over rivalry with other women (aunt, mother) for the affection of the men in the household (father and grandfather). As it turned out, the grandfather became a father surrogate, a position that was invited also by his strength of character. Career choices were carried out with great difficulty, presumably because of the feared consequences of incestuous loving wishes and murderous rage at her female rivals. It was further postulated that a negative oedipal configuration seemed equally probable, in that she was also (through her symptoms and inhibitions) adapting to her anxiety over a homosexual interest in mother, with her father and others the feared rivals whom she might have assumed her primitive wishes had actually destroyed.

Dr. Z.'s masculine qualities suggested unresolved penis envy, which, in fact, emerged, later in the analysis. She felt that it was in a specifically anatomical sense that, by not being male, she had disastrously disappointed her father (as seen by his getting ill and dying). The identification with an older sister was felt to be an additional possible basis for conflict over rivalrous (oedipal) wishes. In this Jewish family where so much stock was set in being successful, the patient felt "unlucky" to be the smartest and yet a female: it seemed to her that success and femaleness could not coexist.

The various losses were assumed to have played a crucial role in her character structure, which seemed phallic-narcissistic ("I don't need you, I don't need anybody"). During the first session she produced a dream in which she was either Copernicus at the center of the universe or lying on a throne that was being carried by runners and upon which she had all the provisions that she would ever need; that is, she was a self-contained system.

It is interesting that during the initial phase of our work the transference reactions went off with great intensity but often too evanescently to be useful or convincing. This early transference instability has been reported by the Parent-Loss Project of the Chicago Institute to be fairly typical of patients with losses during their latency years. Also, when Dr. Z.'s losses emerged, they generally did so in the following order: grandfather, father (postloss), aunt, and father(preloss), that is, in roughly the reverse chronology of their occurrence in her life.

Almost immediately upon finishing her history, Dr. Z. began to experience a fear of falling apart. She thought I must be foolish to have accepted her; did I not know what a sick person she was? I suggested that she, like Groucho, did not want to join a club that would have her for a member. Her response to this interpretation was to remember her mother from the time the mother had a mental breakdown, a few weeks or months after the death of the father. The mother had become profoundly depressed, feeling no one would ever love her again. We began to wonder if Dr. Z.'s experience of her mother's difficulty mourning for the father, as well as her awareness of her mother's Holocaust experience, might have contributed to her tendency toward depressive moods and her expectation of failure in relationships. In other words, the patient appeared to have identified with her depressed mother, who bore the scars of the Holocaust as well as deep narcissistic injury. At times, Dr. Z.'s posture of not needing anyone represented her defense transference, which captured her relationship with the widowed mother (mother's reaction to the father's death had made her distinctly unavailable to the patient) and functioned further as a defense against mourning.

Not unlike what Bergmann and Jucovy (1982) have reported for the children of Holocaust survivors, the patient at her initial presentation portrayed a driven quality and a need to succeed (which Bergmann and Jucovy have postulated represents a compensation for or vindication of the others who could not succeed because they died). In this regard, Dr. Z.

specifically mentioned the sibling who died in infancy in the concentration camp, whom she could easily feel herself to be a replacement for. This came up early in our work, as a speculation, and again later on toward the end of our work, in terms of Dr. Z.'s feeling that she really needed to do more than just be herself, no matter how competent she might be. On the occasions when this feeling would emerge, it seemed easiest (most natural) to extend her own understanding of it by relating it to Bergmann's published experience of other children of Holocaust survivors. This perspective, based on our use of Bergmann's insights, resulted in her being much more relaxed about her creative capabilities, which blossomed as her anxiety started to diminish.

It is of course difficult to capture the adventure of an analysis on paper, just as it would be to catch the excitement of a soaring eagle or a ride down a rapids. I hope the words that follow express some of the intensely meaningful yet complex experience of working with Dr. Z.

During the first year we established mutual trust, which resulted in some insightful collaboration; however, we also dealt with some significant resistances. For example, Dr. Z. would stubbornly refuse to associate at times, because she felt she had "nothing to say." During this phase our work hinged on our mutual ability to communicate affects through many different modes of nonverbal communication. I attempted to put her feelings into words, but she would continue her silence. Yet she could tell from my tone of voice and its intensity (that is, from the musical or prosodic quality of my speech), as well as from my speech content, that I was quite aware of her frustrated internal state and that I only wished to understand it better as soon as she would be able to share it with me. The few communications from her that did occur similarly gave me important clues to her affects through the changing sonority of her voice. Dr. Z.'s voice varied from sad and mournful, to sexual and breathless, to bored and empty, to insightful and intelligent. I tried my best to describe these shadings verbally for her feedback. Of course, at other times it seemed important not to disturb her silences. The judgment regarding when to speak and when to be silent was usually based on my ability to feel connected to her; that is, when I lost the sense of connection (attunement) and had no idea what she was feeling, or why, I would make comments. Although I cannot be sure of it, my impression thinking back over this phase of the analysis is that our mutual sensitivity to nonverbal communication allowed us to maintain a good therapeutic alliance during a painful state in which she was reliving, but not yet ready to verbalize, an important set of disappointing past experiences. These included her mourning for a number of important parental figures and involved aspects of her early identification with her depressed mother. More recent concerns involved her own level of psychopathology.

Dr. Z. retained the fantasy that her previous therapist had rejected her because she was schizophrenic (which was decidedly not the case). It was for this reason that she had started off with profound fears that I would turn her down, either without really giving her a chance to show her stuff or after letting her show too much of "her stuff," which I would diagnose as crazy. She feared I would become disillusioned with her or the reverse—that she would discover some fatal flaw in me. When we explored why she thought she might be schizophrenic, we uncovered the memory of an experience that occurred prior to her psychotherapy during her undergraduate years. Dr. Z. had volunteered to be a control in a schizophrenia research project; she was tested but then not included in the study. No one had ever explained why she had been "rejected," but she had taken this to mean that either she was schizophrenic, and therefore not suitable for being a "control," or that the researcher was not being honest (that is, why else would he exclude data?). She asked and was greatly relieved to learn that she did not show any evidence of schizophrenia, something she had been to afraid to inquire about before this point in the analysis.

We dealt with her pain over past and potential injuries to her self-esteem and her efforts (and my own) to master the analytic method and repeatedly wrestled with her phallic-narcissistic defenses and her early mourning reactions (especially regarding her grandfather). Most importantly, we dealt with enhancing her awareness of her tendency to generally avoid transference feelings entirely. For example, not uncommonly, instead of giving into sad or other feelings she would act out sexually in relationships that would parallel the analysis. These acting out activities I consistently interpreted to her as part of her wishes toward me, with the result that more and more the positive, including sexual, feelings toward me in the transference were experienced as directly relating to me. Dr. Z. next attempted to invite power struggles over almost every subject imaginable: fees, scheduling, freely associating (her need not to at times), and so forth. During this phase my avoidance of entering into power struggles with her often led to her recollection of intense conflicts with her mother. My technique was to be extremely patient with her but to avoid giving in to her unreasonable demands. My facial expression, posture, and tone of voice all communicated my appreciation of and attunement with her agitated state, which lay concealed behind her desperate efforts to manipulate me. Although she lay on the couch throughout the analysis, she would occasionally turn around to look at me to fine tune her understanding of my mood or communication.

At first Dr. Z.'s memories covered the period of her college days. Gradually they moved backward in time to cover her adolescence, her late latency, and finally her relationship with her mother in the period immediately following her father's death and even (at the end of the analysis) prior to the father's death. I will describe this in more detail below.

As an aside and illustrative of our learning, toward the end of the analysis Dr. Z. asked me why I was asking more questions at times whereas at the beginning there were many periods when I had not said anything. I suggested that now I felt I knew more about how to facilitate an analysis and specifically what to ask and when to keep silent. She laughed and confirmed that she also had learned how easy it is to fool ourselves, adding that one of the important things that kept her in treatment during the early painful phases of our work was the belief that although I at times did not seem as competent as I did later on, she basically always had the feeling that I would be honest with her; this, she believed, had helped her become honest with herself and not so focused on presenting a certain front to herself and others. Her mother was seen as especially involved in presenting various fronts, as being very indirect and/or manipulating, an image that never changed during the analysis although Dr. Z. gradually became much more understanding of what her mother must have gone through.

During our second year, along with a continuation of all of the above issues, several new transferences were experienced intensely enough to be convincing. This especially included the grandfather transference and an early father transference. The former was highly positive, the latter negatively tinged. For example, she would become very alert to me, searching me or my room for a particular odor, which she connected with recollections of being with her father. Or she would be struck by some Yiddish expression that would spontaneously emerge either in her own speech or in mine. Sometimes she would comment on the color of my hair, adding gray to it or shifting the perceived color in the direction of her father's hair, as it appeared during his terminal illness. These memories were usually associated first with sadness, later still with genuine pleasure. During this time I recall having a countertransference that was particularly useful in identifying what was going on. I had the passing thought of having surgery for a gall bladder condition, but my main concern was, What would she feel it she ever learned of my hospitalization? Of course, her father had had stomach surgery, and many of the detailed memories of his fatal illness (which came together with powerful affect later in the treatment) especially began to emerge during this phase of our work.

In addition, a negative maternal transference now emerged: she assumed I was ready to boss her around although she clearly recognized that her experience with me suggested that this was not very likely in reality. On one occasion she felt she should get something for my birthday, only shortly thereafter realizing that it was her mother's birthday rather than mine that had just passed. This transference proved to be to a fairly late version of her mother, that is, late in being after the loss of the father. It was from a time in which her mother had become profoundly depleted and had remarried following a period of dramatically lower functioning. During that period the patient felt strongly that the wrong person had died (that is, that it would have been better for mother to have died), but she simultaneously felt relieved that she needed her mother's help in almost all things. There were extremely painful recollections of her mother's state of pathetic helplessness and of her mother's need to have the patient stand in the place of her deceased husband, even to the point of sleeping with her. Dr. Z.'s seeing me as about to boss her around was in the transference a wish that her parents would both be there for her, during her now relived latency years. When she realized this wish, she also remembered a dream from her midlatency in which she was walking across a street and imagined her father and aunt leaning on a balcony in heaven, watching over her. She first had this thought while at summer camp, and it was now the summer vacation that loomed before her. She worried, How could she cope with the loss of me? What if something happened to me? Also during this second year there were transferences relating to the aunt and some significant mourning of her as well. During this break she was able for the first time in her life to visit her father's grave (She did not bring up having visited his grave until I mentioned something in her voice or facial expression that suggested to me that she was sad.)

In the third year we continued the work of the first two years but added themes of intense jealousy, murderous impulses, and guilt (in other words, a full-blown transference neurosis had developed). Positive and negative oedipal transferences began to metamorphose, in varying mixtures. For example, Dr. Z. became aware first of my wedding ring, then of her irritated feelings about my having a wife and a son. (She saw me once with my family in a movie theater and once on the street. On both occasions she saw only me and my son and completely missed seeing my wife (clearly a negative hallucination based on censorship, that is, *not* seeing what *is* there.)

Dr. Z. became aware of her intense rivalry with my family members. As she began working through her feelings about not having me after the manner of her fantasies, she became more devoted to her profession and made significant changes in the depth and quality of her relationships with both men and women. Also, during the second and third year she received recognition for some scientific research. In spite of these positive developments, she continued to hold herself somewhat back in love relationships and still needed ongoing assistance in learning to see her transference reactions when they became displaced onto others outside the analysis, most often involving intense jealousy, envy, or anger.

Gradually but steadily over the third, fourth, and fifth year of analysis, the transference neurosis duplicated more completely aspects of the pre-loss infantile neurosis. There were many dreams in which Dr. Z. would be having sex with her older sister or male cousin (or both) as a disguise for me, where I in turn represented the real grandfather, father, or mother of her childhood. Slips of the tongue (references to her dead mother when she meant father, or vice versa) and her recollection of her failure to even see my wife on the occasions noted earlier began to convince us both that I and my family constituted revenant's for her own family when she was between 5 and 7 years old. She wanted to seduce me, was conscious of feeling sexually excited by me, and would in various ways actually try to accomplish her ends (reminding me at times of Freud's patient who would only accept the logic of soup or the arguments of (real) noodles. Her nonverbal communication made clear that things had come to an impasse, and her frustration grew in intensity. I commented that she felt as though she were "moving deck chairs around the Titanic." She was then able to recognize the connection between her wish to act out sexually with me and her wish to avoid the pain of mourning her father. Still, her resistance became concentrated on one central appeal: that I rescue her by falling in love with her and carrying her off from all wordly cares. She temporaily could not hear my interpretations (given nonetheless) as to why her wishes to make love were understandable, given her history of parental loss and her earlier frustrated wishes to possess her father and mother without rivals. Dr. Z. believed that by not making love to her I was not only willfully rejecting her but "criminally" neglectful! At other time's she would become silent and then berate me for attacking her by withholding (as though I were the silent one). Many interpretations focused on her love for me as a stand-in for one or the other parent and on her sadness over feeling small, insignificant, and unable to measure up. My consistent expression of interest and concern about our properly understanding her terrible predicament of expecting and needing things that could not be enabled us to maintain our alliance throughout this phase, and a number of clarifications regarding the maternal transference now occurred. Dr. Z. frequently assumed that I knew exactly what she was thinking, and therefore when I did not do or say exactly what she wished, I was accused of purposefully frustrating her. When I clarified for her explicitly that such an expectation was based on the assumption that I could and should read her mind, she recalled a period during latency around the time of her father's terminal illness when her mother would read the patient's diary and therefore appeared to always know accurately what the patient was thinking. Dr. Z. partly needed an omnipotent, omniscient mother to deal with the impending loss of her father, who was now recollected (with feelings) for the first time in the analysis as having been precious to the patient.

The patient did not want to hear any more transference interpretations. Yet, somehow, by sticking with them in spite of her protestations, we were able to invite a return of her solid ability to self-reflect, with ever deepening insight. Dr. Z. came to see much more clearly the difference between the current reality and the reality of her transference feelings and was able to make numerous breakthroughs with me to memories of how she had to endure losing her father, who was recalled as her greatest love. She said he had been to her like Alexander the Great. No, he was even better, more like Zeus himself. Or possibly like Achilles, who could not lose after he received the invincible armor that was known as "charisma." She remembered seeing her father in the hospital whereas she had earlier insisted that she had never seen him there because of the hospital rules forbidding visits of children. He had sneezed and she had not said *gezundheit*. He commented on this and when he died a few days later, she had felt responsible, as though this word would have been the magic that was needed to keep him alive.

As the transference neurosis resolved with stunning results, Dr. Z. was able to allow someone to get close to her both sexually and emotionally for the first time in her life, and this relationship eventually matured, after a courtship and engagement period, into marriage. Dr. Z. handled this relationship with new skill and genuine tenderness. She tolerated her fiancé's difficulties without playing into them, and she felt comfortable enough to set a date for marriage. She invited me to attend the wedding. When I told her she needed me more in my role as analyst, she recognized that I could never replace her father, her grandfather, or aunt. Her father had been proud of her; how sad (she now felt) that he could not give her away. After he died, she had been hosted by an uncle of her mother's, and she recalled then feeling her reluctance to attach herself deeply to anyone. This particular uncle and his family had wanted her to stay with them longer, but she had demurred. She seemed to me to be making an important shift toward that side of her family, although I made no interpretations about it. Interpretation was not necessary; an inner, insightful process was becoming self-sustaining. She now appeared to have an appreciation of me, for my genuine qualities, rather than for the idealized image of me she maintained before. She also recalled her anger and even rage at her father's brother, who had not come to her father's funeral for "business reasons." Only now she was able to better understand that he might have had other reasons: like her, he might have been overwhelmed with grief. With this insight and working through she was able for the first time in years to contact her father's brother and even accept some needed financial help from him.

It was time to set a date for termination. Dr. Z. felt as if I had known her longer than she had known her father. She felt that staying in analysis, although pleasant and helpful, would prevent her from fulfilling her personal goals. The attachments to people outside the treatment had exceeded her attachment to me, and she and I finished up our work over a six-month termination phase. She had dreams of sinking in a submarine, of drowning, but now there were indications that the apparatus (ego) was self-directed and self-controlled. There was more anger at her mother and memories of power struggles and fights over such things as what she would wear. In fact, she had not gone to the father's funeral because her mother had dictated a certain dress for the occasion, and the patient had refused. There were other clothes that she had wanted that her mother had not given her. Perhaps, she felt, she was giving up what might have been with me and becoming more accepting of what could not be with her mother; she would have to accept her mother as she is, like it or not.

Having decided to finish and having set a termination date herself, one that made sense to me as well, Dr. Z. began to have doubts about everything all over again. Probably she wanted to hold on to me, to linger in the position of analysand. However, because she continued to operate efficiently in the different spheres of her life and even managed to lose some weight during this period, something very difficult for her to accomplish before, she remained convinced throughout the termination experience that she had made the right decision. But why the vacillation? It was difficult for her to imagine what our relationship would be after the analysis: would I be a friend, confidant, or what? She felt that I would have some indelible place in her memory, but it was also clear that I was no longer the center of her life, as I had been during the analysis. She assured herself that she could come back if necessary, although she recognized this as a vestige of some of her old fears about being abandoned. At the last session she sat for a while, at the end, looking at me face-to-face. She wanted to see me, she said, to make me more real. She asked me some questions about myself and my family, as she had been doing for months. I answered her factually, although I continued my stance of looking for deeper-level meanings in these particular questions; she seemed pleased that we could both analyze the transference and recognize some genuine friendship and mutual respect.

In the year after the analysis ended I received a card, and a year later I received a brief note, which I answered, wishing her well. On each occasion she wrote of her adjustment, which has been successful beyond her dreams. She missed me but with less intensity, and the following year this annual custom of hers was skipped entirely. I conclude from this that she has worked through the remaining feelings about the separation and individuation and that at this point the old conflicts are losing out to her increased functioning in a more conflict-free sphere. She also has the job she always dreamed of, working in a medical research institute that is highly respected, and is living rather than just dreaming some of her more pleasing fantasies.

It has been a deeply moving experience for me to work with Dr. Z. and to watch her mature into a capable and confident woman. The treatment had revived her mourning process and had mobilized an oedipal transference neurosis. As the mourning was completed and the transference neurosis continued to resolve, the arrest in her development was replaced by a series of progressive steps.

The reader can ponder with me the claim made at the beginning of this chapter, as well as in chapter 7 on nonverbal communication, that any psychoanalysis is greatly facilitated by sensitivity on the part of both participants to the nuances of nonverbal as well as verbal communication. My own sense from reviewing the analysis of Dr. Z. is that although her analysis might still have succeeded without attention to nonverbal communication, this aspect of her treatment added a dimension that probably made a significant qualitative difference in the outcome. By most standards such a case would be judged successful, but the level of attunement stayed consistently high, probably because of this communicative connection between the patient and analyst.

10

Psychological Development and the Changing Organization of the Brain[1]

PRÉCIS

A description of the optimal psychoanalytic model should contain sufficient information to explain how the model itself came into being, since the model is of the mind, an "organ" that is in many ways self-replicating. The brain is the kind of self-replicating "machine" that John von Neumann dreamed about and wrote mathematical descriptions of. But few of our developmental models of mind map out the manner in which new mental structure-function comes into being and becomes assimilated into the model itself. Rather, even the best merely describe the series of steps that a particular type of development traverses. Two shifts within science may result in our scientific, psychoanalytic world changing substantially over the next several decades. First, the high-speed digital computer has begun to extend into so-called supercomputer realms, where computers can be used to model behavior of systems of ultracomplexity: the weather, the flow of heat within the mantel of the earth, and the complex activity within the central nervous system (note well, the metabolic activity within the brain is *thinking). Second, we have the combined insights of a large number of sources, which seem themselves to be growing more or less exponentially: economics (decision-making theory), artificial intelligence, learning disabilities science, neuropsychology, neuropsychiatry, psycholinguistics, neurolinguistics, anthropology, archaeology,*

---

[1]This chapter is dedicated to Dr. David Armstrong Breuckner, who pioneered in the area of interdisciplinary research.

linguistics-semiotics, psychiatry, psychoanalysis, social work, communications science, human engineering, robotics, anatomy, physiology, chemistry, neuropsychopharmacology, neuroendocrinology, genetics, neural net research, dream and sleep research, molecular biology, neuroimmunology, theoretical physics (which is getting closer to a "theory of everything"), and ethology. This list is of course partial; the space of this entire book would barely contain a complete list of all such disciplines! In the next 20 years all knowledge relating to the human brain will be programmable into the most sophisticated computer models man has ever seen, which will then be well on their way to learning how to digest, analyze, and comprehend new patterns, and the results will begin to document and validate the small number of remaining viable psychological theories of the brain and demolish those that are outmoded.

If psychoanalysis is to continue to command the respect of the scientific world we must continue to make those kinds of observations and integrations that resulted in the discovery of psychoanalysis in the first place. This means not merely examining our hypotheses and observations to see that they continue to fit with each other (are internally consistent) but also asking new questions that stretch our knowledge at times and reaching out to things more reliable than our current metapsychology. We need to see that our work is not contradicted by the solid discoveries in cognate disciplines. We need to ask more questions: How do things work? How did they get to be this way? Why are they this way and not some other way? Why did change occur in this patient? Why did the patient fail to change? Has the information that the clinical situation provides been thoroughly enough understood and utilized? What are we leaving out? What is wrong with our theory, and what does our theory not explain? Why do only some people learn? What makes learning impossible or difficult for some and like walking downhill for others? Why is learning easier at certain times? Is it possible to change the organization of the brain so as to facilitate learning and if so, in what ways? What really changes within the brains of our analysands or ourselves as we grow in knowledge or wisdom, or lose our minds?

The present chapter is my attempt to push things to the limit and try to set up a series of psychological developmental steps and a series of biological steps and indicate where the two are meant to relate exactly to each other. This, of course, is patently absurd and impossible in the extreme, and the resulting effort is most assuredly wrong in every single particular! Yet my perspective remains that such exercises will be needed as part of the great process of making errors that are really partial solutions and that lead to what is called discovery.

Although few would argue that what is called psychological development occurs as a consequence of changes in the organization of the brain, there are differences of opinion within psychoanalysis regarding whether one needs to understand what these organizational changes consist of. Some believe that psychoanalysis can be practiced without any knowledge of mind/brain correlations whatsoever. However, a counterargument (Trevarthen, 1979; Levin, 1980; Joseph, 1983, personal communication; Kandel, 1983; Reiser, 1984; Basch, 1979, 1985; Gunther, 1987; Schwartz, 1987; Levin, 1988) is that new knowledge of the brain, integrated into psychoanalytic theory, offers a special combination of insight and usefulness that enhances both psychoanalytic theory formation and clinical practice. The validation of scientific theories also requires that they be tested, at times by means of interdisciplinary research.

Development is difficult to comprehend because of its extraordinary complexity. There are many unanswered questions regarding how specific psychological milestones and the early development of the brain relate to each other and are integrated over time. In general, there are also major gaps in our knowledge regarding how basic subsystems of the brain correlate with psychological variables. Moreover, although there are psychoanalytic techniques for getting development "back on track," the efficacy of such techniques will depend upon whether there is accurate knowledge regarding how these measures work. The central section of this paper, with its presentation of specific mind/brain developmental correlations, attempts to address some of the current gaps in our knowledge.

But there are other reasons why it seems especially timely to reevaluate psychological development. Questions about the viability of some psychoanalytic developmental theories have been raised by Demos (1985), Basch (1985), Stern (1985), and Gedo (1989), based upon infant observation studies. Lichtenberg (1983, 1988, 1989) has considered this field of infant observation research an extremely important subject for psychoanalysis. Gedo (1989) has recently presented a revision of his hierarchical model with Goldberg (Gedo and Goldberg, 1973), making the latest neuroscientific understanding of development a linchpin of his theorizing (Gedo, 1989; Levin, 1989). In addition to these psychoanalytic efforts, Adams and Victor (1985) and Trevarthen (1979, 1985) have reviewed infant studies from a neuropsychological perspective, and Meyersberg and Post (1979) have tried to bridge neurology and psychoanalysis with a variety of complex considerations (e.g., by trying out analogies between such phenomena as "kindling," a neurological construct, and "fixation," a psychoanalytic conception). All of these efforts make valuable contributions.

Before we go further, some clarifications seem in order. First, it is most important at this stage of our theorizing to distinguish carefully between the various learning subsystems of the brain mentioned in this essay and the derivatives of such systems, such as self-conscious awareness. For example, no one knows when in life self-consciousness begins, and this paper cannot attempt to solve such a problem. It can consider, however, which learning systems of the brain might be necessary for the eventual experience of a core sense of self. This should become clearer as the discussion proceeds.

Second, Cartesian dualism is assumed to be disproven (see Basch, 1979, pp. 224-226); thus, all psychological variables are considered manifestations of known or at least potentially knowable brain processes, structure, chemistry, and so forth. In addition, in attempting to relate early psychological development to changes in the central nervous system (CNS), there is no attempt to be comprehensive; this would be impossible in a brief essay. Rather, the author's goal has been to select what appears most relevant from a very large body of neuropsychoanalytic research. In addition, because of the confusion that would otherwise result from correlating complex ideas across disciplines, an effort has been made, wherever possible, to differentiate facts from speculations.

Third, this chapter makes certain assumptions about memory and about development. Key parts of the brain are understood to carry essential memory trace circuits, that is, connections that seem necessary and sufficient for learning to occur. According to Thompson (1987), whose orientation is being followed here, the major problem in understanding learning has been in "localizing substrates of learning and memory" (p. 480). In the mammalian brain it appears that "the structures currently thought to be most involved in memory trace formation are the cerebellum, hippocampus, amygdala, and the cerebral cortex" (p. 481). One should add here that the striatum is also important for a second major type of (habitual) learning and memory (Mishkin and Appenzeller, 1987).

The further assumption is made that some other parts of the brain, although not critical for long-term memory storage itself, are nevertheless part of the brain's system for creating and accessing memories. For example, there is evidence that together with the hippocampus, the Basal Nucleus of Meynert is important in enhancing "activation or consolidation" (as well as retrieval) of memories in the neocortex in response to novel visual stimuli that are judged important (Rolls, 1987, p. 531)[2].

As for development, it is assumed that the organizational changes of the brain occur stepwise, in genetically preprogrammed, environmen-

[2]Damage to this Basal Nucleus of Meynert, incidentally, appears to be a critical contributor to the memory deficit in Alzheimer's disease. Clearly, it is not possible to review comprehensively the subject of memory mechanisms within the space of this brief essay.

tally released fashion. Scientists of many persuasions are in the process of specifying what these steps might be. There seems to be a consensus that memory divides functionally into two or more discrete forms, variously described as "semantic"/"episodic," "procedural"/"declarative," or "cognitive"/"habitual." It may help to describe or define these distinctions so that their psychoanalytic significance might better be appreciated. "Semantic" /"episodic" refers to the quality of memory that is either impersonal, left-hemispheric, and language-based (semantic) or personalized and related to the right hemisphere (episodic). "Procedural" and "declarative" are adjectives describing kinds of memories or knowledge, based upon their mode of retrieval. "Procedural" knowledge is elicited through "priming", that is, by action modes or sensory input experience (which, presumably, especially involves the vestibulocerebellar system). In contrast to this, "declarative" knowledge is recalled by direct, language-based requests for information (which appear to tap primarily the corticolimbic system). Finally, "cognitive"/"habitual" refers to specific information-processing systems. Rapidly learned action patterns (habits) seem to involve the corticostriatal system in particular, and in the case of such learning, immature subjects do as well on tests as do adults. This is in contrast to complex, discriminative kinds of learning, which seem to involve the corticolimbic system. This kind of learning requires many repetitions, and adults generally do better than juveniles. (See chapter 6 for a fuller discussion of some of these distinctions and their application and relevance to clinical psychoanalysis).

Finally, I am here implicitly following the interdisciplinary model of the brain described by Ernest W. Kent (1981),[3] which derives from

---

[3]Kent (1981) approaches the brain from the points of view of artificial intelligence theory, psychology, and neurology more or less simultaneously. His model is at once hierarchical and parallel in organization (Kent, 1981, pp. 14-15). There are three major functional systems: input, goal direction, and output. Each of these three systems is represented hierarchically by low-, intermediate-, and high-level processors and interconnections with the other systems at each level. Data paths are multiple and parallel (often simultaneous). As one proceeds toward higher levels, one moves from large numbers of perceptual receptors and internal state indicators towards smaller networks of integrators and synthesizers. The output of the motor system is into the external world (which is also represented within the CNS), and the actual external world becomes a critical input to the sensory apparatus. The sensory analysis system (i.e., input system) and the motor (i.e., output) system are connected by the intermediate system, which measures internal states and determines which goals are to be targeted within an individual's goal hierarchy. The vast majority of decisions are assumed not to require conscious decision making and never reach higher levels of organization. A much smaller number of decisions are routed to the higher and

Ludwig von Bertalanffy's general systems theory. Kent's model is clearly an information-processing model. Basch (1973) has pointed out that Freud himself was extremely close to developing such a model in his "Project for a Scientific Psychology" (Freud, 1895) although he never published this during his lifetime, and in fact focused his theorizing instead on an explicit model based on the distribution of libidinal energy, rather than of information. Basch (1985), Rubinstein (1973), and many others have emphasized the current importance of the information-processing perspective for psychoanalytic theorizing. This perspective is employed in this paper as an intervening variable, that is, as a crucial step in the process of constructing a linkage between psychoanalytic and neuroscientific conceptions, to make the integration of concepts less of a speculative enterprise.[4]

Because of time and space limitations, it will not be possible to review what is known generally about the development of the CNS per se. Those interested should consult Yakovlev and Lecours (1967), Adams and Victor (1985), or Gilles, Leviton, and Dooling (1983).

SPECIFIC PSYCHOLOGICAL, DEVELOPMENTAL STEPS: THEIR SEQUENCE AND POSSIBLE NEUROLOGICAL CORRELATES

In a series of research efforts, I (Levin, 1989a, in press; see also chapter 2) have tried to build a case for combined neuropsychoanalytic models. In what follows, the focus will be on the evolution of learning-related subsystems of the brain that eventually become the basis for a core sense of self, for the cohesiveness of this sense of self, for certain psychological defenses (including the formation of a repression barrier), and for control over sexual and aggressive drives. These psychological developments are presumed to be the consequence of the following specific changes in brain organization:

(1) It is known that from the first days of life there is an actively functioning cerebellum, which will ultimately connect with the cortex via

---

highest levels of the goal selection system, where an extremely small number will actually enter consciousness. It should be clear that sensory analysis and coding of data play a critical role in coding for meaning, and possibly an additional role in establishing a kind of "machine language" that the brain can use to communicate with itself (see Levin, 1988b).

[4]An example of this approach is my analysis of connections between dream and sleep research (see chapter 6). In that chapter, sleep and dream research are each first examined in terms of learning subsystems of the brain and then related (on this basis) to each other.

the upper brain stem and thalamic structures. Recent PET scanning research has shown (Chugani and Phelps, 1986) that from birth onward the cerebellum (and also the thalamus) is extremely active metabolically and is therefore in a position to contribute decisively to "early" memory systems, subserving especially postural and coordinative motor control.[5] *It is worth speculating that this cerebellar (memory) system might serve as the basis for the first self-related experience, what might be called the core sense of self.* These archaic memories (chiefly of motor experience but possibly also involving other sensory modalities) in the newborn and young child would be loaded into the cerebellum (the most primitive motor-memory system that we know of) and potentially transferable later on in development to higher centers (e.g., in the brain stem, cerebral cortex, etc.) when these subsystems would become optimally functional. The result would be linkage and mutual enhancement of the various systems mentioned. Consistent with Piaget (Piaget and Inhelder, 1969) and the most recent theorizing of Daniel Stern (1985), this concept of the core self would thus be based on the earliest sensorimotor (cerebellar) memories, available from birth or very shortly thereafter, as a coordinating influence within the brain.

(2) It is known that the cerebellar memory system eventually becomes the vestibulocerebellar system (VCS), which unfolds as a bidirectional control loop (the VCS-cortical system), with consequences for the coordination of action initiated at the cortical level (see chapter 3). *It is speculated that this coordination by the VCS-cortical system is an important step in the establishment of early physical self-cohesiveness.* See below regarding clarification of the word "cohesiveness" in the present context.) This cohesiveness would then improve and ultimately be expressed

---

[5]According to Chugani and Phelps (1986, p. 841), the cerebellum is active at birth and even decreases in activity during the first year of life. Also, by the age of 3 or 4 months the parietal integration zone becomes especially active and, along with the cerebellar hemispheres, aids " visuomotor integration." The basal ganglia and limbic system mature at 3 months, the frontal and associative cortices at about 7.5 to 8 months.

A separate question is whether what is true for processing visual information is also true for other modalities. One cannot be sure, but I believe this may be correct. For example, support is found in recent work on the visual strip that shows plasticity similar to that found by Merzenich et al. (1984) (see chapter 3) in the sensory-motor strip. Kaas et al. (1990) show that "cortical neurons that normally have receptive fields in [an experimentally lesioned region of] the retina acquire new receptive fields in portions of the retina surrounding the lesions" (p. 229). This research is complex and therefore difficult to interpret, but it is suggestive that the adaptive mechanisms are similar regarding differing sensory modalities (in this case vision and tactile sensibility).

along with a variety of other self-related functions (agency, coordination, the early mapping of the sensory experience, awareness of body parts, their relationship to each other, and their position in space, etc.). But the increasing cohesiveness would be the result of the simultaneous activity of all self-related memory systems (cerebellar, striatal, cortical, limbic, etc.).

At this point it may help to define the word cohesiveness as used in this context. Usually this word refers psychoanalytically to "resistance to fragmentation," that is, the resistance against losing the sense of self-sameness, even in the face of significant emotional stress. In the present discussion of the word, the usual psychoanalytic meaning is still intended; however, a more strictly biological sense is included as well, namely, that property of memory that resists disruption so that what is learned can be retrieved in a timely manner (whether habit pattern or discriminative learning) and used for adaptive purposes. Clearly, the implication of using these two meanings together is that a relationship is postulated to exist between a psychological experience on one hand and a biological function on the other. Presumably this resistance to memory interference is experienced as an internal stability of self; moreover, any resistance to memory interference would be a consequence of the processing of critical, self-related long-term memories within multiple memory systems, each making its own contribution to the momentary sense of self and serving as a check on the possibility of failure of memory retrieval within any particular subsystem. Put another way, the multiplicity of functional biological systems for the processing of self-related information provides "backup" (i.e., reliability), much as does the redundancy built into certain strategically important physical systems (e.g., the computers on board a space shuttle).

(3) It is known that the central parietal cortex, which has unique abilities in the area of integrating experiences across the different sensory modalities, develops a map of the self in an absolute space, that is, according to a coordinate system, (Kent, 1981). *It is speculated that this maturation of the central parietal cortex (and other crossed sensory integration zones) contributes, possibly decisively, to the cohesiveness of our early self-related memory systems.* This would be the functional or psychological significance of what Luria called the "tertiary cortical system."

At least two other structures also make important contributions of the mapping of the self in space, the first being the hippocampus (Rolls, 1987). "The hippocampus is ideally placed anatomically . . . in that it receives highly processed information from association areas such as the parietal cortex (conveying information about position in space), the inferior temporal visual cortex (conveying a visual specification of an object), and the superior temporal cortex (conveying an auditory specification of [the same] stimulus)" (Rolls, 1987, p. 526).

The second additional structure contributing to self-mapping is the amygdala, which specifically accomplishes this by means of rich sensory input. "It is possible that the amygdala not only enables sensory events to develop emotional associations but also enables emotion to shape perception and the storage [and retrieval] of memories" (Mishkin and Appenzeller, 1987, p. 88). The mechanism for the learning thus described apparently involves (as a necessary stage) the release of endorphins, opium-like neurotransmitters, which are present within the neuronic projections of the amygdala (p. 88).

Before we continue, it will help to consider briefly the function of crossed sensory integration, as is accomplished by such structures as the central parietal cortex, the amygdala, and, presumably, other brain structures as well. By whatever unknown mechanism this synthesis occurs, the result is that ongoing experience becomes coded independently of the original sensory mode(s) involved. Put another way, this step in information processing takes one from a sensory (or primary) stage of memory, in which various cortical and other (sensory) maps of experience are created, through a secondary or short-term stage, to a tertiary stage in which long-term memories have now formed. These are distributed in some unknown form, that is, the knowledge contained in the original experience is now coded in some "machine" language[6] of the brain; it is "abstract" (see Kent, 1981). Clearly, the unknown process involved in the tertiary stage is of fundamental importance for information processing and knowledge acquisition.

(4) *It is known* that over a period of years a system of two properly functioning cerebral hemispheres with a high level of interhemispheric

[6]An additional fascinating sidelight to this discussion, touching on the role of sensory experience, is the observation by the late David Marr (1978, p. 165) that when the brain processes a sensory modality, such as vision (and other modalities likely follow this pattern as well), and creates (a series of) internal visual maps, these maps become progressively refined (abstracted) and also more and more oriented toward the visual perspective of the experiencing subject. This suggests that the abstracting phase of long-term memory (LTM) is more individually variable than the sensory phase of memory. That is, as we process information and go from the sensory to the LTM storage stage, we make progressively more arbitrary choices about how to store particular knowledge. It is possible that the meaning of all experience is more idiosyncratic than we might ever imagine. From this perspective it seems logical that the "filing" code of LTM is probably highly personal. Apparently, these biases show up as quickly as experience gets recorded.

(i.e., left-right and right-left) connectedness comes into being. *It is speculated that the resulting integrative tendency in cognitive/affective processing that results from the integration of the two cerebral hemispheres*[7] *makes a further contribution to cohesiveness and to the early formation of the repression barrier.*[8] As the cortical ramifications increase, involving complex interplays between the cortex and lower centers (striatum, thalamus, cerebellum, etc.), more subtle and still more abstracted self experiences become conceivable. As noted earlier, our actual sense of whom we experience ourselves to be would seem to be a kind of derivative of all of the self-mapping, self-abstracting memory systems (central parietal, limbic/ hippocampal/amygdalar, thalamic, striatal, cerebellar, etc.). The richness of the product would seem to flow from the multiplicity of specific memory systems involved and their unique contribution to the overarching system.

(5) *With the further refinement of the system of right cerebral hemisphere and limbic system* (which have intimate connection with each other [see chapter 3]) it is known that affects are better regulated, and *it seems worth speculating that what psychoanalysts call the repression barrier further matures on this basis.*[9]

(6) *It is speculated that the remainder of the development of this defensive function, which Freud called the repression barrier, is accomplished by the increasing and reversible dominance of the left over the right hemisphere,* which is known to occur during brain maturation (Flor-Henry, 1983; personal communication, 1986). That is, the assumption of left-hemispheric dominance provides us with improved control over

[7]The subject of how the hemispheres are integrated is a complex, unanswered question, beyond the scope of the present discussion. That aspect of the subject regarding the defenses of repression and disavowal, and the cerebellum in particular, is discussed in chapters 2 and 3.

[8]A fuller discussion of the neurophysiology of "repression" by Hadley (1987, pp. 3-4) points out that in addition to the more advanced form of (secondary) repression representing left-right interhemispheric blocking, there is probably a "primary repression" based upon the storage of engrams in basal areas and the cerebellum, memories that never reach consciousness (p. 4). She further refers to a kind of "passive repression" based on information storage in the right hemisphere, which needs left-hemisphere mediation to become accessible to consciousness; this latter information can, however, have access to "the motivational system through the basal ganglia and nucleus accumbens" (p. 4).

[9]Steps (4) and (5) depend upon the evolution and refinement of a large number of changes, including the establishment of specific psychological defenses, that the author believes involve specific neurophysiological mechanisms, as pointed out previously regarding the examples of "repression" and "disavowal" (see footnotes 7 and 8).

sexual and aggressive impulses (Flor-Henry, 1983; personal communication, 1986).

(7) If the preceding brain developments (together with other developments unknown or too complex to consider at present) eventually come together to form an overarching system with unique, "efficient" characteristics, then experience becomes possible at the end of the spectrum that has been called the "conflict-free sphere" or the area of "transformations of narcissism" (Levin, 1973).

THE CONTRIBUTION TO COHESIVENESS OF THE VESTIBULOCEREBELLAR SYSTEM

Recent research on autism is establishing that the cerebellum is indeed a crucial element in the learning process (Hadley, 1989, personal communication). (Chapter 3 brought together in one place a number of important observations that relate to the cerebellum and its role in our mental functioning.)

Whatever our sense of self is, it would seem to include as a necessary element the body/mind self, that is, some representation of the body parts and their relationships to each other, and some representation of the world of other people and of the inanimate world. In addition, all of these "things" eventually become organized by our brains within a coordinate system of absolute space. There are those (Kent, 1981) who feel that it is, in fact, the placement of our self-representation within such an absolute coordinate system that constitutes a decisive addition to the crystallization of self.

The self-definitional functions mentioned above are all properties of the vestibulocerebellar system (VCS). In addition, the VCS is active in coordinating the hemispheres before the corpus callosum becomes myelinated. It follows naturally that what psychoanalysts call the self, which involves self-definition, differentiation, and coordination (see Gedo, 1989a), in all likelihood requires an intact cerebellar database. Whatever happens during psychosis or fragmentation states would seem to involve at least some perturbation in the availability of information from this cerebellar database or within other core areas responsible for crossed sensory integration.[10]

But there is still another aspect of the cerebellum that suggests an important role in early self development. There is reason to believe that within the cerebellum we create a model of the self-in-the-world and that our manipulation of thoughts about ourselves and others (i.e., our think-

[10]In this regard please see the discussion, pp. 197–198, regarding Crow's work on schizophrenia and amygdalar malfunction.

ing) at one time occurs by means of experimental manipulations or adjustments within this cerebellar model (Itoh, 1984a, 1985a). Without this cerebellar model to "play with", it is very likely that one would need actually to manipulate objects just to think about them! It follows, then, that some individual psychopathology that appears overly instinctual and primitive in terms, for example, of the need to touch—or mouth—may actually be a reflection of an arrested psychological development that correlates best neurologically with the expression of an incomplete or primitive cerebellar system (model) for cognition in the absence of any compensatory or more advanced system.[11]

To put these points somewhat differently, our ability to predict follows from the brain's ability to create internal models of sensory experience (Kohonen, 1983). To this end there exist multiple, organized, sensory maps within the brain, such as the somatosensory parietal map (see Lusek and Kaas, 1986, p. 83), tonotopic maps in the temporal cortex, and maps within the visual cortex, the cerebellum, thalamus, hippocampus, brain stem, and so forth. These feature maps are aspects of the self organization of the brain and lead to further abstractions about the self and the world we live in (Kent, 1981, p. 265). But they are maps without which there is no mapping; that is, they are part of the fundamental machinery necessary for information processing (and knowledge acquisition) itself.

THE TWO CEREBRAL HEMISPHERES

To the extent that the two cerebral hemispheres represent processors of our ongoing experience, whose collaboration is required for the optimal learning subsystem of the brain to be brought to bear on a particular problem, the intact hemispheres working together by learned and hardwired means reflects a major quality of who we are. Our very identity is connected, much more than we realize, with the cognitive/ affective style of this interhemispheric collaboration. The goals, qualities, and limita-

---

[11]According to Hadley (1987), the Kluver-Bucy Syndrome (resulting from the removal of portions of both temporal lobes, including the amygdala on each side), which produces a condition in which the subject has an excessive need, among other things, to touch or to mouth objects, may well be "the operational equivalent of throwing the organism back from limbic or cortical control of behavior into 'cerebellar mode' " (p. 3). That is, in the early cerebellar mode, before we have created a self-in-the-world model, our learning/thinking requires that we touch everything in order to cognate. Since children are building up such a model, they generally need to touch and/or mouth everything, that is, such behavior is required for their internal model construction.

tions of each hemisphere are unique. What may change during either analysis (or normal maturation) is the ability of the two hemispheres to complement each other. Creativity and self-reflection may also be consequences of more interactive use of the two cerebral hemispheres.

A final word about the hemispheres. There is reason for believing that there are many specific disconnection states wherein the hemispheres fail to adequately communicate, resulting in a self that is significantly divided. We have speculated that some of the so-called defenses, such as disavowal and repression, may be nothing more than the manifestation of disconnections in different directions (see chapter 2). Although we have implicated the VCS as a likely participant in such "disconnect" decision making, it should be obvious that much more needs to be done to understand psychological defenses in terms of brain physiology. Also, the implications for psychoanalysis of these mind/brain correlations are only just beginning to be appreciated (Levin, 1989).

THE CORTICOLIMBIC SYSTEM

The corticolimbic system makes major contributions to the emerging and ongoing sense of self, as indicated in the preceding discussion. Pioneering work on the limbic part of this system by Dax, Brocca, Papez, MacLean, and Nauta has been reviewed by Nieuwenhuys (1985). Central to this work has been the "triune" brain theory of MacLean (MacLean, 1960, 1985; Moore, 1988). Of the three-brains-in-one, the first, or reptilian, brain subsumes programs for basic physiological need satisfaction; the third brain (the neocortex) receives sensory input about the external and internal worlds, from which it generates a composite image or "double exposure" (Moore, 1988, p. 6). The second, or lower mammalian, brain (which MacLean named the limbic system) plays a key role in social and emotional behavior, including language. It is through the limbic brain's contribution to successful bonding with the mother and others that the survival of the individual and the species is assured.

In humans and other mammals, the right hemisphere has a particularly close affiliation developmentally with the limbic system. Thus, the right hemisphere also plays a critical role in the identification of faces and the management of affect, two functions that therefore appear to be closely related.

MacLean's pioneering work, as noted previously, helped establish some of the particular mind/brain, developmental correlations that were presented earlier in this essay. Two additional areas of research on the corticolimbic system, however, have decisively influenced my thinking and will be noted briefly at this point.

T. J. Crow (1985, 1986a, b, 1987, personal communication) has demonstrated that in right-handed identical twins, there is a clear concordance between schizophrenia and a defect in the amygdalar portion of the left temporal lobe, as demonstrated on CAT scanning. Reynolds (1987), working on autopsy material from the brains of the same patients, has shown that these damaged amygdalas contain abnormally high amounts of dopamine. The significance of these findings is that it is now possible to explain, in at least one class of schizophrenic patients, that the symptoms are due to damage to a single area of the brain subserving both affect and cognition. Put differently, on the basis of damage to part of the limbic system, the anhedonia and cognitive deficit of this disease become more readily understandable, and the critical role of the left amygdala in mental functioning becomes more obvious.

There is also evidence that the dominant hemisphere is essential not only for the control over the motor system, but also, for most of us (Flor-Henry, 1983), for a left over right hemispheric control of affects as well. This relates particularly to major affects that are of interest to psychoanalysis, namely, sexual and aggressive drives, which Flor-Henry connects principally with the right hemisphere. Flor-Henry (1986) has collected evidence from male exhibitionists, for example, that identifies a defect in part of the left temporal pole (identical in all patients thus far studied), as seen on computerized EEG recordings. From this and other evidence he has concluded that the loss of control of these individuals seems best explained, physiologically, as a failure of the left to "police" the right hemisphere.

Hadley (1985) has also stated, regarding the amygdalar and hippocampal systems for information processing (the systems disturbed in both Crow's and Flor-Henry's patients), that psychological hatching is likely "a behavioral correlate of the maturational coordination of multiple sensory, motor, attentional and affective processes" such as is contributed to by the amygdala in particular (p. 536; also see pp. 537-41). Clearly, one can investigate such processes from either an ontogenetic or phylogenetic perspective. The former perspective represents psychoanalysis; the latter, the biological perspective. This paper has attempted to relate the two by means of a developmental series with mind/brain correlations.

COMMENTS ABOUT THE OVERALL SYSTEM

As discussed in chapter 3, the overall (and oversimplified) system of right hemisphere/limbic system, left hemisphere/motor system, together with critical brain stem nuclei and the vestibular cerebellar system, comes together to form a unique brain both in mammals generally and in *Homo sapiens* in particular, with very special learning capabilities. The human brain, under optimal organization, is capable of being self-reflective, coherent, and cohesive under a vast array of stresses; it also appears to be self-organizing as a fundamental property. Moreover, our brains are particularly capable of learning within a relationship with the brain of another member of our species in a one-to-one relationship, whether it be parent/child, teacher/student, therapist/patient, or friend/friend. Whether we are correct or not in our ascription ascribing many roles to the cerebellum, including an important function in learning, this observation about learning in dyads would still appear to need an explanation.

In trying to better understand development this chapter has focused on research that offers the possibility of dividing the brain into parsimonious functional units. Some will differ over what these units should be; of course, to some extent this decision is arbitrary. Those interested in more of the relevant neuroscientific research might wish to consult Nieuwenhuys (1985), Kety (1982), Shashoua (1982), Reinis and Goldman (1982), and Agranoff (1978). Obviously, the brain's considerable complexity continues to create problems for all of us in attempting to fathom how best to bridge psychoanalytic psychology and neuroscience.

SUMMARY

Eventually we will know both the psychologically relevant parts of the brain and their schedule for development. In this paper the following correlations and sequence of development have been presented (as a speculative enterprise):

1. Immediate postpartum availability of the cerebellum (VCS) as a critical contributor to the formation of archaic, self-related memories (making possible, eventually, the beginning of a core sense of self)

2. Establishment of VCS/cortical system as a bidirectional control loop providing tentative "cohesiveness" for this ancient memory system

3. Maturation of the central parietal cortical system (as well as the hippocampus and amygdala bilaterally), adding polymodal and self-mapping potential and thus decisively augmenting "cohesion"

4. Development of two properly functioning hemispheres, their connectedness and coordination, resulting in further solidity to the self-related memory systems, and early formation of the repression barrier (based upon functional interhemispheric blocking as early psychological "defense")

5. Preferential connection between the right cerebral hemisphere and the limbic system, with resultant improved regulation of affects and a strengthening of the repression barrier

6. The establishment of dominance of the left (or language-related hemisphere) over the right hemisphere, with a resulting consolidation of the repression barrier and more effective control over the "drives"

The system presented in roughly tripartite terms coincides well with the analysis of a number of other theoreticians, for example, the system described by Mesulam (1981, p. 314), who considers a system wherein an interacting sensory component (i.e., posterior parietal cortex), exploratory/motor component system (i.e., frontal cortex), and motivational map (i.e., the cingulate cortex) feed into and are fed by a system for arousal, attention, and vigilance (the reticular structures, such as limbic connections, nucleus locus ceruleus, and brain stem raphe nuclei). In addition, Mesulam's views are clearly compatible with Kent's model as described earlier in this chapter. The point here is that there is sufficient agreement among scholars that it is time to consider in more detail a bridging of psychoanalytic psychology and neurology. We know that the sensory input into our brains relays to unimodal and eventually polymodal areas of the cortex (see chapter 1) and that various abstractions that psychoanalysis is concerned with then become possible. These would include the very sense of self at the core of our being, our sense of self-cohesiveness, our psychological defenses, our repression barrier, and our control over sexual and aggressive drives. To change this core, to understand how and why our psychoanalytic technique works, we might at least not exclude a priori detailed knowledge of the brain; and neuroscientists would do well not to ignore the insights of psychoanalysis.

11

Some Notes on the Evolution of Language

PRÉCIS

One may argue that the evolution of language is of only marginal interest or relevance to psychoanalysis. And perhaps this is correct. But I believe that psychoanalytic theory, especially regarding psychological development, needs to take into account the emergence of language; that is, the theory must not conflict with what is known to be true about language and its evolution. The real problem is that we are very uncertain what language is in terms of brain functional systems. For example, many people who are deaf use AMESLAN, a syntactical sign language that involves visuomotor or spatial perceptual skill (presumably right-hemispheric, since the right hemisphere is where such skills are usually concentrated). Yet when deaf people suffer strokes of the left hemisphere (the usual hemisphere for syntactical language in hearing people), they develop aphasia or language disturbances, just as do hearing people! This is especially interesting since it means that the left hemisphere is really a syntactical language hemisphere (for both hearing and deaf people, even though the languages of these two groups seem to be based on different principles) and that visuospatial functions used for sign language and those used for other purposes are different brain modules with different locations within the brain. Obviously there is a need for further research to delineate the principles upon which languages are based.

A second point can be made about the relevance of language evolution to psychoanalysis. Syntactical language is an example of a phenomenon that is uniquely human, and therefore our understanding of the details of its development should help us understand the relationship between nature and nurture, that is, how genetic and environmental factors interact to produce complex developmental steps.

Of course, in general terms syntactical language capability is a consequence of the evolution of an enlarged and more complex human brain and of neotony, the relative immaturity of this brain at birth. Since human development is delayed or incomplete at birth, there is dependency on adult caretakers. This dependency necessitates the prolonged exposure to adults that, together with man's genetic programming, determines who he shall become (Gould 1977, p. 401–404). In the final section of this chapter, I wrestle with major diverging views about the relative importance of experience and genetics especially as they might relate to questions about the evolution of language.

Knowledgeable readers will observe, however, that the current chapter does not systematically present the perspective of semiotics, although some such theory does enter into the discussion, for example, when considering the contributions of Saussure (to Chomsky's theorizing) and in the mention of work on primate language by Premack and others. Those interested in a general review of semiotics might consult Innis (1985). Nevertheless, a brief review of this subject seems appropriate.

According to Innis, modern semiotic research begins with C. S. Pierce, F. de Saussure, and K. Bühler, who established the foundation of the field by considering and delineating the details of the so-called sign function of language (Innis, 1985, pp. viii–x). Aristotle's conception of language as based on metonymy and metaphor is comparable to later semiotic conceptions of "contiguity" and "similarity" (see chapter 7), and the sign function of language is also captured in the Scholastic formula "aliquid stat pro aliquo" (something standing for something else) (Innis, P. 1). Semiotics, also called semiology, includes linguistics as a subcategory and is the science that studies "the life of signs within society" (Saussure, 1959, p. 28). Language, according to these pioneering researchers, covers multiple fundamental functions: appellative, conative, expressive, and representational. For example, Bühler states that any psychological theory of perception is rendered meaningless if the semiotic perspective is ignored (Innis, p. 67).

At the beginning, a great deal of effort went into understanding the mechanisms of representation. Jakobson sees Bühler's work as most seminal in this regard. Bühler (1982) in turn credits H. Gomperz with the most thorough working out of the theory of representation and its implications, based on the idea of the capturing in one object of "accidents" that normally inhere in another object, thus establishing a "relation of meaning, of standing in place of, or representing" ["Bedeutens, Vertretens oder Repräsentierens"] *(p. 73).*

Jakobson (1960) gets credit for significantly expanding the explicitly considered functions of language to include emotive, referential, poetic, phatic, conative, and metalingual aspects (p. 154). In a similar manner Bateson (1955) deserves credit for describing the paradoxical, playful, and creative aspects of language. And Langer (1942, 1967) must be mentioned for her description of the fundamental differences between discursive and presentational symbolism (see chapter 1). What is important for our discussion of language is that these theoreticians, especially Bühler, see language as a complex, multilayered set of functions that provide communicative relatedness (within a society or group) of various sorts and at various levels of abstraction. In somewhat different words, the communication of humans and other animals allows interaction and cooperation within each particular species. Bühler (1982) comments specifically on the role of the "schreckruf" (or cry of fright) wherein animals warn others of their group about the presence of ominous smells or sights of danger (p. 71). An evolutionary perspective thus proves helpful in appreciating human total communication, and with this view semiotic knowledge concurs.

What specifically does semiotics say about the relationship between genetically predetermined language capacity and the actual skill individuals acquire in their native language usage? The answer seems rooted in Saussure's conception of "langue" and "parole," which becomes a core conception of Jakobson, Chomsky, and others. To appreciate this core concept properly, we should "listen" momentarily to Saussure (1959) himself:

> *We first singled out two parts [of language]: language and speaking. Language is speech less speaking. It is the whole set of linguistic habits which allows an individual to understand and be understood . . . But this definition still leaves language outside its social context; it makes language something artificial since it includes only the individual part of reality;* [but] for the realization of language, a community of speakers ["mass parlante"] is necessary. *Contrary to all appearances, language never exists apart from the social fact, for it is a semiological phenomenon. Its social nature is one of its inner characteristics. Its complete definition confronts us with two inseparable entities [p. 45; emphasis added].*

That is, language is a theoretical function, but it is also inseparably bound to the existence of a community of speakers. For Saussure, actual usage ("parole") stimulates language development in the individual and even results in the gradual change of the particular native language itself over time, providing a combination of continuity and change and "varying degrees of shift in the relationship between the signified and the signifiers" (p. 46). Saussure's perspective is remarkably in contrast to Chomsky's, which is monolithically genetic and anti-experiential.

The answer to our question is, therefore, that nature and nurture are complexly interactive, and this helps us appreciate language development both ontogenetically and phylogenetically. Human language is really total communication in that it involves multiple modes (functions) of communication, which operate more or less simultaneously. From the perspective of semiotics, the multiple functions of language are provided by the multiple subsystems for language-related behavior. And from the perspective of the phylogeny of language, human syntactical language does not replace other language modes already in existence but becomes added to the mix of possibilities. That is, human language in the larger sense is an outgrowth or extension of the language of earlier mammalian and simian life-forms. Bateson and Hinde have both commented on the complex relationship between verbal and nonverbal communication (Sebeok, 1977, p.300), as have I in chapter 7. What remains is for us to use this knowledge of semiotics to better appreciate the nuances of language growth in individuals and in the species.

It is possible that a detailed study of the various language modes or categories I and semiotic specialists have observed will help us considerably with the details of language evolution. Specifically, it may be that each language subsystem has not only its own profile of hemispheric localization (see chapter 7) but its own specific rules for acquisition and its own degree of genetic loading. The individual study of each of these different gestural, prosodic, semantic, and syntactical systems will result in a more complete description of how or whether these modes interact and how much their development in the species and the individual is autonomous or requires "priming" or "releasing" mechanisms from the group. It is also possible that these different communicative modes have different windows of time for their optimal development and different mechanisms or potentials for recreating such windows at later dates. When language is looked at semiotically, and not as a monolithic Chomskian deep structure of the brain, there seems to be more hope of increasing our understanding of how language evolves.[1]

[1]The role of inheritance in behavior has recently been reviewed by Polmin (1990), who writes that "unlike single Mendelian characteristics, genetic variance for behavioral dimensions and disorders rarely accounts for more than half of the phenotypic variance, and multiple genes with small effects appear to be involved rather than one or two major genes" (p. 183). Therefore, nongenetic factors contributing significantly to the overall variance need to be identified, along with their specific mechanisms involving such developmental phenomena as age-to-age

In the future, it seems that zoosemiotics (Sebeok, 1977), the special application of semiotics to animal communication, offers an especially useful perspective in solving such problems.

I am using the following definition of language: that internal brain system which can either be used (1) "in order [deliberately] to influence the behavior of other animals" (Wells, 1987, p. 1), or (2) for the purpose of private rumination. The major benefits of language relate to its allo- and autoplastic functions (that is, its utility in altering the outer, human world or ourselves). Human language, from a biological perspective, would appear to be the consequence of a limited number of definable evolutionary steps supported by changes inside and outside the brain. The following summary account draws on information from vertebrate evolutionary biology, archaeology and anthropology, linguistic studies, and neuropsychology. Some parts of the story seem clear; other parts require speculations that will probably prove correct; and still others involve significant leaps of imagination and should be accepted cautiously.[2]

An overview starts with a diagram of the evolutionary line from which mankind eventually emerges. This line includes agnatha, placodermi, chondrichthyses, osteichthyes, amphibia, reptilia, and mammalia (in this order). After the reptile stage was reached, an additional 150 million years were needed for the transformation "from small, cold-blooded . . . reptiles to tiny, warm-blooded, furry mammals" (Carrol, 1988, p. 362). Then, about 200 million years ago, the first mammals appeared on earth, after a preliminary stage of mammal-like reptiles called therapsids. One of the decisive changes in the therapsid stage from the perspective of verbal-vocal language evolution was the appearance of sensitive hearing (that is, hearing adapted for the reception of sounds in the higher frequency range) in association with a new kind of infant–mother bonding (compared to the behavior of reptiles) wherein the infant's vocalizations could supplement other communicative signals between the infant and

---

change, continuity of genetic effects, and the role of environmental factors in gene activation or inhibition (see Polmin, p. 187).

[2]Hoyt Alverson, a linguistic anthropologist, introduced me to the work of Hagman (1982), which I highly recommend for the reader seriously interested in holistic approaches to language. Hagman attempts to integrate the various biopsychosocial aspects of language and notes the many subtleties and pitfalls involved. His major point is that to understand the evolution of language one must appreciate its utility to the human species. The expanded definition of language offered at the beginning of this chapter is influenced by the views of Alverson and Hagman.

the mother (MacLean, 1985). Changes in the vocal and hearing apparatus reached a peak with the evolution of mammals.

Mammalian evolution involved major transitions: "The entire skeleton was modified, as were the soft anatomy, behavior, and physiology down to the level of cellular metabolism" (Carrol, 1988, p. 361). These changes are a complex story that deserves a separate description. From the perspective of language evolution, however, the important step of *improved hearing* becomes possible because of the appearance of three delicate but effective inner ear ossicles: the malleus, incus, and stapes. These bones, together with the advanced mammalian tympanum (eardrum), qualitatively increase the sensitivity of the sound impedance matching system far above what is possible for reptiles.

Reptiles will eat their young if they can find them. Obviously, it is adaptive that reptile babies are unable to vocalize (MacLean, 1985, see chapter 10). Generally speaking, "sounds are less salient in most reptiles," and it is hard to tell by means of vocalization how reptiles feel, since they do not cry, whine, or scream when hurt or suffering (Burghardt, 1988, p. 126). In contrast to this, therapsid (and all subsequent mammalian) mothers incorporated changes in the thalamostriate division of the limbic system that programmed them to (1) want to nurture their infant offspring (MacLean, 1985), and (2) vocalize for the purpose of expressing affect clearly. With this crucial (mammalian) change in both motivation and vocalization, development shifts imperceptively toward the human form of prosodic (rhythmical and emotional) communication.

Summarizing briefly, what is decisive for human language evolution in the reptile/early mammal transition are three shifts (based on the work of MacLean, (1985): (1) Unlike reptiles and earlier life-forms, early mammals and their descendants organized experience (and memory) around the principle of *affects* (MacLean, 1962). We now take this for granted, but it was obviously not always so (in our reptile and earlier "ancestors"). The adaptive advantage of this shift is enormous, since being able to classify and store experience along such a dimension as pleasure-pain helps us avoid what is dangerous and seek what is helpful. (2) There is a shift from olfactory to *auditory and visual orientation*, and this is part of what leads to the increasing importance of such nonverbal communication as gestures, facial expression, sonorities (intonations), and so forth (MacLean, 1962). (3) And there is, as noted, a fundamental shift in *the form of attachment* between mother and infants, and this (together with neotony) is what will eventually allow for the acquisition of syntactical language, partly on the basis of the infant's prolonged exposure to the caregiving adult's language.

The story of the evolution of more complex life-forms is also the tale of a general shift from genetically preprogrammed systems of instinctive knowledge, available from birth, to a more pivotal role for experience-based systems of knowledge. For example, although some fairly complex learning is possible at the level of reptiles, much is nevertheless preprogrammed into the organism "instinctively," relative to mammals (Burghardt, 1988, p. 126).

Primates appear at the very end of the Cretaceous period, about 65 million years ago. Anthropoids begin in the lower Oligocene period in Africa about 38 million years ago. Beginning 22 million years ago hominoids appear, *Proconsul africanus* being the earliest of the group ancestral to both the later apes and humans (Carrol, 1988, p. 473). Between 4 and 14 million years ago the earliest hominids come into existence, although fossils are largely missing from this part of the record. Molecular evidence based on DNA comparisons suggests that man and the great apes diverge during this period, roughly between 6 and 10 million years ago (p. 474). Then about 4 million years ago hominid fossils begin to appear in the form of *Australopithecus*, the oldest of the genus being *A. aferensis* ("Lucy"). The evidence suggests that *A. aferensis* possessed hands that were more precise at manipulation than those of the chimpanzee (p. 474). This is important because it underscores the probability that *gesturing* was the major form of communication.[3] *Australopithecus* walked erect, hunted, and used stone tools of a simple nature (Washburn and Harding, 1975, pp. 6–8). The comments that follow regarding these human ancestors are surely open to considerable debate.

Starting two to three million years ago, at least three lineages coexist: *A. aferensis, A. boisei,* and *H. habilis* (the last, especially, seen after 1.75 millions years ago). One million years ago *H. erectus* spreads out of Africa into eastern and southern Asia. *H. erectus* hunts large animals, uses fire, and creates symmetrical tools by a technology requiring months of practice (Washburn and Harding, 1975, p. 8). Three hundred thousand years ago *H. sapiens* essentially replaces *H. erectus* (p. 475). *H. sapiens* is anatomically modern, and the archaeological record contains his shelters, graves, art, and a prolifery of tool varieties (p. 8).

By 100,000 years ago *H. sapiens* evolves into his modern form, living alongside Neanderthal man from 70,000 to 30,000 years ago. The important question for this narrative is obviously: When does gesturing give way to more total communication, which would include standardized

[3]Freud himself (1923) commented that "thinking in pictures is, therefore, only a very incomplete form of becoming conscious. In some way, too, it stands nearer to unconscious processes than does thinking in words, and it is unquestionably older than the latter both ontogenetically and phylogenetically" (p. 21).

vocal or verbal communication? Some speculations about this development appear below, but the consensus is that speech (vocalization) generally increased in our ancestors as brain size increased (Washburn and Avis, 1958).

Although hardly an expert in this field, the evidence related to the onset of language is discussed by Leakey (1981) and consists essentially of four kinds of data: (1) studies of primates who have been taught sign language; (2) investigations of brain imprints within fossil skulls; (3) the study of ancient stone tools; and (4) the broader archaeological record, which includes art objects, cave paintings, and other material evidence of ancient rituals (e.g., burials).

Primate studies have been inconclusive regarding language evolution. Some scholars believe that apes cannot really use sign language in the manner of a formal language. These investigators focus on the residual differences between modern primates and man. However, some of us familiar with sign language are not inclined to quibble over exactly what constitutes "formal syntactical language." This latter group believes the great apes and chimpanzees *can* use sign language for emotionally meaningful and even at times creative communication (Premack, 1981). The problem, however, is that whatever position one takes on the highly controversial issue of whether primates can "sign," it matters little in determining an answer to our question about language evolution. The reason for this is that the primates being studied are not in fact ancient primates at all. They are no less advanced than we are and are likely as different from their ancient ancestors as we are from our own. Therefore, even if scientists could agree that modern apes and chimpanzees can use signs or sign language, this does not prove that the ancestors whom we shared with them 6 to 10 million years ago had any such ability. It will be seen, however, from the following discussion that there is nevertheless a strong probability that the gestural communication capacity of our apelike ancestors did play a role in human language evolution.

The imprint of the brain on fossil skulls has been studied extensively by Holloway at Columbia University, who has concluded that "the basic shape of the human brain is evident in hominids of *at least* 2 million years ago" and can be seen as well in the smaller brain of australopithicines from 3 million years ago (Leakey, 1981, pp. 131–132). As for language, *H. habilis* skulls are said to show a clear impression of Broca's (motor speech) area of the brain. The Broca's area impression in *H. habilis* is also allegedly more prominent than in the great apes but not as prominent as in modern humans (p. 133). In comparison, *H. erectus* has an even more prominent Broca's area. But Leakey and Holloway feel that this information, although suggestive, is not sufficient to allow one to conclude that verbal language was present in these early hominids. A safer conclusion would seem to be that these early hominids were most certainly capable of uttering and hearing *vocalizations* because all known primates can do this. The intonations or sonorities of these utterances would have conveyed meaning (for example, about affect states) to self as well as others. Nevertheless, this vocal capability should not be construed to mean that verbal communication had developed yet, nor that vocalization was even the most important communicative mode employed. Nonverbal gestural communication, as is used among the primates, might well have been employed (Goodall and Hamburg, 1975), along with some vocalizations (Washburn and Harding, 1975), for a long period of time prior to the appearance of standardized systems of vocal or verbal communication. Put somewhat differently, in the transition from apelike ancestor to man changes need to occur in the pharynx (it must lengthen) and in the tongue (which requires a more posterior position) in order to produce the formant frequencies and the fricative, nasal, and other noises that make up the 40 or so human phonemes (Young, 1978, p. 186). Also gestural (nonverbal) communication needs to evolve more fully into a language system, and changes need to occur in the brain that will permit the eventual emergence of symbolic capacities.

My position that syntactical language is built on a nonverbal language foundation has a long and distinguished history. Its highest expression, however, occurred during the Enlightenment when E. B. Condillac, Thomas Reid, and J. B. (Lord) Monboddo explicated man's invention of language. As Wells (1987) describes, these three (especially Condillac—and later Jean-Jacques Rousseau) state the argument in detail: (1) in the beginning, signs (gestures) were in all likelihood not intended as communications but rather were merely "normal reactions to particular situations" (Wells 1987, p. 8) that later were discovered to have communicative-adaptive value; (2) sonority undoubtedly played a role in communicating affects, although vocalizations would have been clarified by accompanying gestures, and (3) context would have further helped in decisively clarifying the meanings(s) involved in the overall communication. Unfortunately, these insights of Condillac and others were not properly understood, and in fact were severely criticized by influential men such as J. G. von Herder, and subsequently became unpopular (pp. 31–47). One might add that a secondary reason for criticism would have been an antiscientific religious bias against such an evolutionary conception. So, in place of Condillac's well-thought-out theorizing, what prevailed is Herder's orthodox view that human language is unique unto itself and that no transition from gesture and prosody to syntactical language seems possible (see Wells, 1987, especially p. 35). As we shall see later, Chomsky's perspective contains a measure of Herder in it.

Leakey (1981) reports especially on Glynn Isaac's work on the link between tools and language. The archaeological record of tools begins about 2 1/2 million years ago with the beautiful teardrop-shaped hand axes of *H. erectus*, the so-called Acheulean industry (p. 134). *H. habilis* apparently had tools approximately 2 million years ago (Alverson, personal communication). Stone tools before this time (questionably associated with the early australopithicines) apparently show neither symmetry, complexity, nor a sharp sense of purpose (p. 135). Then 100,000 to 200,000 years ago a new method of tool making, that of striking flakes, the Levalloisian technique, appears. This method is more economical still and might be a marker for further change in communicative mode. From this time until 40,000 years ago the Mousterian industries appear, in which the Levalloisian technique flourishes in the context of the first signs of burials, grave offerings, and cult objects (p. 135). After 40,000 years ago there apparently is a cultural explosion in terms of variations, artistry, and so forth.

Studies focusing on tool use in chimpanzees do not appear to be contributory to our understanding of tool behavior and its correlates in man. In addition, no chimpanzee has been able to use one tool to make another (Goodall and Hamburg, 1975, p. 18).

According to Leakey (1981), Isaac interprets the evolution of stone tools as indicative of behavior associated with increasingly complex rule systems in which sequential processes became paramount (p. 136). If we remind ourselves that sequencing is critical for all language processing (see Levin, 1989b; also see chapter 10, this volume), we will have to agree with Isaac and Leakey (as well as Washburn and Harding, 1975) that Mousterian tools and other evidence of early man supports the conclusion that by 40,000 years ago human *verbal language* is fully evolved. The change from Acheulean industry to Levalloisian technique suggests that something significant also happens regarding language evolution about 100,000 to 200,000 years ago (see the following paragraphs).

The evidence from the archaeology of art is more complex than that from stone tools, as one might expect. However, the overall thrust of the material corroborates the same explosive degree of change occurring after 40,000 years ago, or possibly a bit earlier (Leakey, 1981). Painting, ancient sculpture, and other artifacts demonstrate the symbolic and nonutilitarian nature of early man's created-object world. This would seem to be prima facia evidence of an increasing ability to abstract; this ability, together with the ability to sequence (seen from tool analysis), makes extremely likely the coexistence of developed verbal language, although this remains a speculation.

A clarification may be in order. Leakey's (1981) conclusion (p. 138) that a rudimentary form of verbal communication arose as long ago as 2 million years (the time of *H. habilis* and the australopithicines) most likely refers to an intonational communication system of utterances (vocalization), as noted earlier. Differing from Leaky, my own belief is, however, that the first language system was primarily based on gestures, aided by some nonstandardized utterances. Quite complex and refined communication can (and still does) occur by means of gestures alone (see chapter 8).

The appearance of *H. erectus* then leads to more complex communications, as evidenced by the appearance of ritual acts with symbolic significance. What possibly happens with *H. erectus* is that the gesture system becomes significantly more standardized (more like a syntactical language) and therefore more serviceable for communication of belief systems (and other kinds of information) while, at the same time, the verbal-utterance mode lags behind gestures in development.

This description regarding the shift in language mode to something more like syntactical language is necessarily vague because no one really knows how the nonverbal communication system that we are born with (and that is functional from birth) dovetails with the inborn syntactical capability of man, which develops during a slightly later window of time.

A best guess about what happens next would be that about 100,000 to 200,000 years ago, intonation (or sonority) becomes combined more systematically with the already well-developed gestural-language foundation that man shares with his simian ancestors (Hewes, 1973; Lowenthal 1982; Vygotsky, 1934, p. 35). Within all post-reptilians intonation plays a role in communicating affect, but the question remains: In what way do vocalization and gesture now become mixed?

My reason for placing the gestural system ahead of the verbal system in terms of language evolution is twofold: (1) there is evidence that modern human beings do not usually perceive consciously the operation of their own or another's nonverbal communications, and I belive this means that the older gestural system has already become automatized in a way that verbal communication has not been; (2) there is evidence that man's capacity for nonverbal manual communication developed significantly before his laryngeal (speaking) apparatus matured sufficiently to create refined vocalizations (Lieberman, 1984). Hopefully, the reader appreciates that these comments represent an initial attempt to organize an extremely complex subject.

The basic idea underlying this description of changes that led to man's contemporary total communication system is that at each step the mutative changes in some subgroup or individual results in adaptive advantages and increased survivability. For example, the systematization of vocalization or verbalization would be expected to improve cooperation in group activities, such as hunting, warfare, child rearing, and food preparation.

Finally, about 40,000 years ago the intonation apparatus becomes much more developed and vocalizations more standardized. This would lead readily to the beginning of the use of vocalizations for denotation (naming)[4] as we know it. Once naming occurs, a series of complex, abstract developments would be possible for the individual and the group that would lead to the rapid evolution of formal verbal syntactical language with the naming of increasingly abstract entities.

What is crucial about naming is that it allows for the storing of information according to a tag that has nothing to do with the original sensory aspect of the experience itself. Of course, language is much more complex than attaching names to sensory experience. At some point, meanings become elaborated for phenomena of increasing abstractedness, as language development progresses. I have described this evolution elsewhere (see chapters 8 and 10) but believe that the essence of the early phase of language development (abstraction) can be simply described in the following manner: a name-sound is connected with each of the major sensory modalities of an experience and therefore can serve as a way of relating these aspects of an experience to each other. The sensory experience then "falls out" and what is left is the "name" as a mnemonic for the "abstract quality" that the sensory experiences share. D. Stern (1985) has described aspects of this abstracting process under the rubric of "amodal perception," which he believes to be the basis for the infant's experience of a unified self and object. This is turn leads to attunement and object ties. Premack (1981) believes that he has demonstrated conclusively that primates are capable of "amodal perception".

To summarize, human language derives from the evolution of mechanisms that first support nonverbal communication in animal behavior and that pass on to both man and modern primates from their shared ancestors. Before the arrival of therapsids (early mammals) hearing in the air (compared with hearing in the water) plays only a minor role because the middle ear mechanisms for hearing in the air have not evolved. Reptiles communicate by means other than sounds, and their affects, which are not an organizing principle of experience, are not apparent vocally. In other words, the critical changes (secondary to programming

[4]By "naming" I am telescopically describing a complex process involving multiple developmental steps and not assuming a simplistic one-step process. As Litowitz and Litowitz (1977) state, "It is an adultocentric view of early child language that children are only 'naming' whereas a complex, dual process is involved in which the two planes of language each become manifest; the associative (that is, vocabulary) and the syntagmatic (that is, syntax) " (p. 430). In chapter 7 these same aspects are referred to as the principles of similarity and contiguity and are discussed in relation to the work of Ferdinand De Saussure, Roman Jakobson, and Jacques Lacan.

shifts in the thalamostriate division of the limbic system) involve (1) the development of hearing in air in association with the signal of vocalization (the use of sonority), (2) the use of affect as an organizing principle for memory, and (3) the bonding of mothers and infants. Finally, once bonding became a factor, vocalization could become of adaptive advantage, since it would help therapsid mothers and babies bond. If reptile mothers had heard vocalization from their babies, in contrast, they would only have found it easier to locate and eat them!

At first, intonation carries part of the message, and gestures carry the other part. Primates and all other mammals are adept at employing both gestures and intonations, more or less simultaneously, and probably neither system was standardized in earliest man (*H. habilis;* Acheulean industry; 2 million years ago). But in the case of *H. sapiens* the *intonation (sonority) system becomes especially important for carrying the affective message, and verbal language develops on top of (but never completely replaces) the older, nonverbal, gestural system.* First, the gestural system gradually became standardized (*H. erectus,* 1 million years ago). Then the standardization of intonations/sonorities also occurred gradually (*H. sapiens;* Levalloisian technique; 100,000 to 200,000 years ago). Finally, in a creative development, the "naming" process was discovered and proved of adaptive advantage (*H. sapiens;* 40,000 years ago). What proved decisive was man's long prior experience with nonverbal communication and with a small basic vocabulary of affective sonorous messages. These evolved over eons of time, the product of subtle mutations within the brain, skull, jaw, middle ear, and so forth, with increasing adaptive advantage always coming from improved communication and cooperation within the species. It therefore seems plausible that *communication grows from the integration of nonverbal and verbal communicative modes. Human verbal language follows from the evolution of the integration of humanoid hearing, vision, and touch sensations (which I also contend is an important part of the basis of our ability to abstract).* And out of this intersensory integration (especially once "naming" is discovered) there rapidly develops the human capacity for advanced psychological development, culture, and syntactical language itself, more or less as we know it today.

How vocalization and verbalization systems become integrated, as well as how syntactical communication modes become integrated with nonverbal communication into a system of total communication, remains to be delineated in detail. It seems that no matter how far we come, there remains a residuum of unanswered questions to ponder. Because no archaeological or physical evidence is available to enable us to piece together the final steps in the story of language evolution, the reader is left to chose between a variety of possible alternatives.

One scenario, quite reminiscent of Freud's account in "Totem and

Taboo"—but without any attribution to Freud—is given by Gans (1981), who sees the decisive event in terms of Rene Girard's theory. According to Gans and Girard, language (gestural or verbal) emergence involves a "context" or shared "event" and a "universal signifier" (pp. 11–15). The event is assumed to be a murder, carried out by the primal group and then inaugurated as a ritual event, thus giving the victim's sacrificial body a complex of sacred meanings (pp. 8–12). My major objection to Gans's work is that he sees no need to integrate his ideas with the insights of others working in closely related disciplines. He develops his viewpoints essentially as a closed system. For example, although he quotes Derrida (regarding Derrida's idea of "presence") and Lèvy-Strauss, there is none of the ecumenism of these scholars in Gans's thinking. And in Gans's need to exclude philosophy and biology he naturally also leaves out Derrida , who may well have originated the idea of "context" (a concept with a definite relationship or even equivalence to "presence"). In addition there seems to be no reason for assuming a priori, as Gans does, that a sacred murder within the primal horde is more likely to be central to language evolution than any other communal event.

A second set of polarities about language needs to be considered: the debate between Piaget and Chomsky over empiricism versus "innatism" (Piattelli-Pamarini, 1981). This may represent a modern echo of the older controversy between Poincairé and Russell at the beginning of this century questioning whether knowledge is intuitive or logical. To keep within the boundaries of this paper, however, we will concentrate on that part of the disagreement between Piaget and Chomsky that relates to language and, most specifically, that bears on the question of how language evolves. Technically, neither Piaget nor Chomsky comments on this specific question; however, their unique approaches have proved so influential within neuropsychology that I cannot adequately present my interdisciplinary perspective about the birth of language (ontogenetically and phylogenetically) without contrasting it with the theorizing of Chomsky and Piaget.

Chomsky's view (1972, 1975) is that language represents a deep (organlike) structure of the brain with its own innate but discoverable rules of operation. His emphasis on sentences (syntax) rather than words (phonemes) revolutionized the field of linguistics, although it should be noted that modern structural linguistics derives as well from Saussure, the Copenhagen school (Hjemslev), the Prague school (Troubeskoy), and the French school (Martinet), each with its "emphasis on language as a formal and abstract system" (Litowitz and Litowitz, 1977, p. 424). Thus, abstraction ("langue") differs from actual utterance ("parole") to use Saussure's terminology. Chomsky's transformational rules are merely a rephrasing of Saussure's insight and stand for what connects the so-called deep structure on one hand and surface structure on the other; that is, the representational system for language and the actual language performance, respectively (Litowitz and Litowitz, 1977, p. 439).

Given the possible derivation of Chomsky's concept of transformational "grammar" from Saussure's distinction between language as abstraction and as performance, it is especially interesting that Chomsky does not see language development in the individual as representing anything other than the unfolding of the genetic blueprint for language; that is, he sees individual experience as an essentially trivial factor in development (as compared to the predetermined genetic plan). His perspective, like von Herder's, regarding human (versus primate) communication is that human language involves a unique system, unlike any other in the animal kingdom (for an alternative view, see Lieberman, 1984).

Chomsky is singularly uninterested in the research of computer scientists, such as Papert, or primatologists, such as Premack (see Piatelli-Palmarini, 1981), whose research deals with important neurophysiological aspects of language closely related to Chomsky's own work. He seems totally focused on the narrow problem of decoding or delineating principles (rules of transformation) that he hypothesizes to be a universal language grammar. It seems especially ironic that although Chomsky insists that any proposed universal principle of grammar be stated in a manner that is falsifiable, he does not consider the implications of having chosen a hypothesis (that of a universal grammar) that he himself admits is not itself falsifiable (Piatelli-Palmarini, 1981). That is, Chomsky will, by definition (that is, by choice), always remain in the position of being able to tell others that their work is not germane to his own, while never having to really prove that his most fundamental postulate is correct!

On the positive side, however, it should be noted that by his serious attempt to study the logical, deductive, mathematical properties of syntactical language rules and his debunking of Skinner's "verbal behavior," Chomsky greatly stimulated the field of developmental linguistics (Litowitz and Litowitz, 1977, p. 428). It is also possible that the form of language is, as Chomsky surmises, predominantly determined by genetics. I cannot help but wonder, however, why Chomsky remains so apparently uninterested in any other perspective than the genetic and why he is so closed to examining the role for environment-genetic interaction. It may be possible that although Chomsky takes the position he does formally, he is not as wedded to deduction, nor lacking in an interest in the empirical, as he seems.

Piaget's theorizing allegedly starts from the observation of children's development. One could also argue, however, that the observations themselves flow from Piaget's own conceptual framework of stages of cognitive development. While Chomsky considers language as something present in man, *ab initio*, Piaget (1971) conceptualizes steps of sensorimotor schema (up to age 2) leading to concrete operations, preoperational thinking, and, ultimately, abstract thinking (around the age of 12). Piaget spent his lifetime studying children, recording and comparing their responses and attempting to integrate this mass of data on a variety of cognitive issues into an epistemology of cognition. He is no less convinced than Chomsky of the correctness of his approach, although he tends to downplay the role of genetic development in favor of experience. Perhaps the greatest difference between these two theoreticians is their varying appreciation of the role of individual experience: Piaget feels that activity on the part of children sets the stage for change and permits the actualization of built-in potentials; Chomsky seems supremely skeptical that experience might effect any change.

How then do my views of the evolution of language dovetail with these two different frameworks? I have stated that human language appears to be a complex achievement, with man merely demonstrating his place in the animal kingdom by means of multiple modes of communication (used by his ancestors at various stages of evolution). I have delineated the various evolutionary steps that lead to gestural and prosodic systems and have suggested that these nonverbal communications constitute a language foundation with which mankind starts (just as every human infant starts out life with these same gestural and prosodic systems, available from the very beginning of life for communication with others). Later, through millions of years, various stages of systematization occurred so that eventually systematic gestural and verbal (that is, syntactical) languages evolved within the human species. However, man continues to communicate within multiple simultaneous communication or language modes. Modern apes (for example, chimpanzees) appear to share man's potential in this regard. I have also clearly suggested that religious and other anti-evolutionary prejudices originally interfered with acceptance of the work of those scholars of the Enlightenment (such as Condillac) who first stated this theory of language evolution.

At first glance it appears that the theory of language evolution presented in this chapter is most harmonious with Piaget's thinking and at odds with that of Chomsky. It is certainly correct that Chomsky would be singularly unimpressed with such a theory, since his sense is that man is unique and that, therefore, man's language is also. Piaget's position of stages of accommodation and assimilation would seem more consonant with the theory of language evolution, positing as it does that each step builds on an adaptive accomplishment of the preceding stage.

However, I would like to state for the record that I do not see my theory as consistent with either a Chomskian, primarily genetic theory or a Piagetian experiential perspective. This is because neither of these theories about cognition adequately explains the interaction between hereditary and environmental (that is, experiential) factors.[5] It would be safest to admit that we are still in search of a theory that properly explains the interrelationship between these variables. In other words, it seems likely that Piaget's stages of cognition would not be generally universal if genetic factors did not, as Chomsky asserts, play an important role. However, to say that genetics is decisive is to overlook, as I believe Chomsky does, all evidence to the contrary relating to the role of experience. This includes important linguistic research, such as the experience with deaf children of hearing parents who (more often than the deaf children of deaf parents) are likely to fail to obtain adequate or optimal language stimulation. It is known that a significant number of children within this group will therefore never develop proper syntactical language (see Chapter 7). This can only mean that during a window of time determined by our genes the language program unfolds, but only if the proper environmental experience obtains.[6] To put this into Saussurian-Chomskian phraseology, "parole" (performance) seems to prime the development of "langue" (the abstract language system). In terms of the discussion elsewhere in this book, the language function seems to be like those other learning subsystems of the brain that require priming for optimal development. It is my contention that such priming is vastly overlooked.

The point is that psychoanalysts and psychiatrists working with the deaf have shown repeatedly that language development is just one more area in which experience and endowment meet head on. Scientists of different persuasions are urgently needed to help develop new paradigms that can better delineate the complex relationships between experience and endowment. Although Hubel and Weisel have done so for aspects of the visual system, the complexity of the brain requires that the other sensory and learning-related systems be independently studied.

---

[5]Wolff (1967), among analytic contributors, makes a similar assertion that "neither an environmentalist conception of imitation and conditioning, nor an apriorist conception of inborn faculties [alone], conforms to the facts of language learning" (pp. 301-302) in which children, through exposure to native language, intuit its particular syntactical rules and acquire the ability to speak.

[6]In chapters 2 and 3, through a consideration of the phenomenon of learning, especially learning readiness and levels of brain plasticity, I have approached the problem of how nature and nurture interrelate. Writing on similar adaptive learning mechanisms in nonhuman primates, Washburn, Hamburg, and Bishop (1974) point out that development "is first a period of protected learning" during which youngsters' play involves the practice and mastery of survival skills (p. 7). In fact, if any species were unable to adapt, that is, "if they were totally bound [by instinctual patterns] to the past, they would [rapidly] become extinct" (p. 7).

Although a new paradigm is certainly not yet at hand, I would like to mention my own intuition here. It seems that an important clue to better appreciating the interaction of nature and nurture in language development would be the probability (referred to elsewhere in this book) that certain key learning subsystems of the brain serve multiple functions. Such subsystems can become crucial knowledge bases, but they may also serve as leading parts of the brain's system for total communication by allowing one or another communicative mode. As argued in this chapter (as well as in Chapter 7), man communicates by means of multiple, simultaneous (parallel) communicative modes. It would be interesting to learn to what extent these modes might also provide linkage between different parts of the brain itself. The prosodic, gestural, semantic, and syntactical language modes are the product of a long evolutionary history and may be the means to understanding better how the language of the brain and native or natural language connect with each other.

12

Overview

The title *Mapping the Mind* was chosen because the idea of mapping connotes the optimal activity for a stage of exploration in a field with relatively few landmarks. Such is the state of interdisciplinary research correlating mind and brain. It has been important for the many scientists involved to carefully delineate meaningful psychological or functional units and their possible neurophysiological correlates so as to guide our thinking and future research intelligently. Science grows by incremental steps in which old and new theories are compared by experimentation, the results communicated to colleagues, and periodic shifts made in paradigms when enough new evidence is accumulated to seriously question older theories.

I hope that in this book I have accurately conveyed the pioneering studies of the individuals mentioned, along with my personal sense of excitement about man's voyage into the terra incognita of the human brain. Our brains are as novel a territory to us as the discovery of the New World was to the citizens of late 15th-and early 16th-century Europe. The explorer Amerigo Vespucci wrote in 1503 to Lorenzo de'Medici about a "new world" (S. Schwartz, 1980, p. 14). Similarly, the current generation of research in psychoanalysis and neuroscience is creating a radical opportunity for mankind to improve its own conditions on the basis of the possibility of a united knowledge of mind and brain. Until we have such a unified theory, however, we will need the assistance of maps to help us navigate through the complex interdisciplinary perspectives involved. Thus, the purpose of this book: to conceptualize and map out the general dimensions of mind and brain, as they are currently available, but at a level of specificity and detail sufficient to be maximally useful to scholars and clinicians alike.

There is much in the world that is both arbitrary yet practical. When the 15th-century mapmaker Waldseemüller named the New World in honor of Amerigo Vespucci, placing America approximately where Brazil stands today, the word became popular so quickly that he was unable to retract it when he later changed his mind about who most deserved commemoration in this manner! As Einstein noted (see chapter 9), the exact connection between sensory experience and concept formation is also arbitrary yet practical, more analogous to the relationship of check number to overcoat than soup to beef.

Many formulations and speculations make up this book, but one of the more central ideas is that *the brain's primary function in ordering experience involves some arbitrary, complex, yet eminently practical, neural control mechanisms,* with feedforward and feedback qualities. For example, the stream of sensory input is transduced into appropriate output (actions or inactions involving adaptive decision making and judgment) based on a hierarchy of goals and values and information about inner state (including affect). To this end, the prefrontal cortex selectively utilizes from the rest of the brain a system for the pursuit of its goals and further directs the selective attention that guides the sensory system so that principally "desired" (meaning appropriate and useful) sensory input is obtained. Most simply stated, what we see determines what we want, but what we want, know, and are capable of conceptualizing also impacts upon what we look for, see, and appreciate. In an interesting parallel to this interactive relationship between sensation and cognition, it is also true that there is an interactive relationship between neuroscience and psychoanalysis. Although neuroscience can help psychoanalysis appreciate the basic design features of the brain, without psychoanalysis the optimal appreciation of the significance of this biological data will never occur.

A second important concept about the brain discussed in this book is that *the brain's core organization seems to rely on a series of learning modules or subsystems that develop over time and that also function as knowledge bases and communication modes.* Formal language exposure ("priming") plays a crucial role in neuropsychological development, decisively altering brain organization in a direction that allows for abstract thought and its consequences (native language acquisition, advanced psychological development, and the capacity to experience and contribute to culture). Nonverbal communication, including mime and other gestures and the ability to communicate through vocalized intonation (sonority), represents communicative modes present from birth (that is, genetically programmed). It is on this foundation of nonverbal communication that formal language is ultimately built (both ontogenetically and, probably, phylogenetically as well).

In somewhat different words, *it is possible that what we call synthetical, or native, language is the exploitation for interpersonal benefit of a language originally used by the brain for internal communications.* At the very least, native language and the operating system of the brain share rules (grammar). This important point is discussed in chapters 4 and 5 in relation to the research efforts around the world to understand such major illnesses as schizophrenia. Prominent in this work is the research of Ingvar (1988, 1989), Niwa (1989), Hiramatsu et al. (1982), and Crow (1984, 1986a, b). A third core concept is that it is increasingly possible to specify *how learning and development occur as reflections of the capture of experience within the various levels of the brain's "plasticity"* (which have a relationship to the learning-knowledge-communicative mode subsystems mentioned earlier). The unfolding process commonly called psychological development, as well as the means by which we choose to learn, appear to be genetically programmed. But the overall process, its success or failure, nevertheless appears to be contingent on critical and timely environmental factors (often meaning the beneficial input of parental figures). The adequacy of this input is predictable in the species but random for individuals. Some of the many ways learning and development become arrested or skewed have been explored clinically and theoretically in the text, but I hope the reader appreciates the view that *various subsystems of the brain seem to leave "fingerprints" or characteristic "dialects" in communication that help us identify their participation in various mental-emotional processes.* The awareness of these indicators on the part of psychoanalysts would seem helpful clinically, leading to improved recognition of the nature of learning difficulties (that is, as Hartmann pointed out, functions such as learning will be established with either primary or secondary autonomy). For example, sometimes normal neurophysiology of defenses is involved (that is, the problem is "psychological"), and the recognition of this will result in the analyst's serving in his usual capacity as a translator between different parts of the mind or brain, allowing output from one aspect of the self (ego) to become input for another aspect, thus facilitating learning. This will differ, however, from a situation where the analyst senses that thinking itself is focally unusual, with the pattern being more a neurocognitive difficulty rather than a psychotic thought disturbance or neurotic, conflict-based, problem. Still another example would be appreciating when priming might be helpful in order to restore procedural memories that form part of an important transference complex, thus serving as part of the process of helping the patient overcome the repression of specific content and the disavowal of the meaning of important past experience.

Summarizing these points briefly, I have specified how transference interpretations accomplish a variety of effects, including the following: the bridging of the hemispheres (or other subsystems of the brain); stimulating cerebellar-based archaic memories; interpreting cortico-striatal habit patterns acquired preoedipally (representing acting-in or acting-out); tapping lexical systems within the dominant hemisphere; identifying gestural and/or prosodic systems bilaterally to clarify what the current relevant affect(s) might be; and identifying the still different "amusicality" that means neurocognitive deficit.

This brings us to a fourth core concept: *the process we call abstracting may be the result of the integration of sensory input across modalities,* which at a higher level of activity within the brain would involve the consolidation and "downloading" (transfer) of information within the three or more major knowledge bases of the brain. The resultant or product is assumed to be a master image that has added dimensionality or depth of field, an achievement with survival (adaptive) value.

In discussing intersensory integration and distinguishing it from synesthesia and amodal perception, I have been led toward two theories: one about the transference phenomenon itself and the other about REM/nonREM cycling. It seems possible that REM/nonREM cycles might be understood in terms of consolidation and "downloading" phases within a process of data acquisition from sensory input. Assuming there is some truth to this conjecture, I have further posited that transference and nontransference phases might characterize waking states much as REM and nonREM cycles characterize sleep. By connecting sleep and dream research in this way I am seeking a neurophysiological basis for Freud's insight that dream life deals with the expression of wishes and factors that impede wish fulfillment. My tentative conclusion is that within the basic organization of the brain (whether asleep or awake) *there is a strategy to deal with unresolved wishes and what opposes wish fulfillment by creating a compromise product that expresses both opposing tendencies.* This would represent the familiar psychoanalytic concept of symptom or compromise formation, and from a neuroscientific perspective it would represent a more basic strategy of the brain. This more basic strategy is the pattern noted earlier *to create internal knowledge bases with increasing "depth" by means of combining the slightly varying perspectives of different subsystems.* To better understand what I mean, consider how depth perception itself is created by combining the slightly different binocular sensory input into a single master image. I am merely suggesting that this strategy might be used more generally within the brain.

A fifth core concept is that *the prefrontal cortex, basal ganglia, and cerebellum are the crucial elements in the brain's operating system and this neural control provides the matrix not only for communication with others but within the brain itself.* The exciting work of Ingvar, Niwa, and of Tsunoda (chapters 4 and 5) suggest possibilities about how *the brain's language and the mind's language seem related to each other* and how these instructions might be rearranged for given tasks (for example, coded into the messages of our native language).

Finally, transference as a phenomenon has been discussed as something that is most likely adaptive for the species. Transference behavior is met with either complementary or noncomplementary responses. The former will allow frustrated archaic needs to be met. The latter will result in further frustration but also an opportunity to compare the experiences within the transference state with those when there is no transference. These different (contrasting) perspectives can provide a basis for an expanded view, that is, one with extra depth. This means that even the negative responses to transference can potentially result in personal insights about oneself and about relationships. Most important, however, transference is seen as a vehicle for the communication of affect, and to this extent a special plea is made for analysts to become knowledgeable in the various nonverbal, as well as verbal, modes of human communication. *Language is not verbal communication alone but something far more complex: it is a system of multiple communicative modes* that includes facial expressiveness, hand and body gestures, posturing and mime, the prosodic or musical aspect of verbalization, and the syntactical systems that we call native language. A great deal remains to be done to clarify how such subsystems are handled by the brain, that is, what their physiological substrates are. To my thinking, Freud made major contributions to the study of language through his development of *psychoanalysis,* which *is a system par excellence for decoding the meaning of human communication.* If Freud's unique semiotic perspectives continue to be applied along with others (see chapter 11) we will learn much more about how mind is an outgrowth of brain and how human feelings alter the brain in fundamental ways.[1]

[1]One additional point can be usefully made, namely, that current neuropsychological research supports the idea that motor activity and adaptive learning are fundamentally associated with each other. Jeannerod (1985) reviews the relevant research in this area, including prominently the work of Hebb, Held, Hubel and Weisel, and Piaget. An interesting question is why adaptive learning does not occur unless the subject initiates motor actions. Jeannerod believes that the need for self-initiated (not reflexive) motor action is a consequence of the fact that in a sense every motor movement is a test of an hypothesis about the self in the world. Thus, successes and failures in planned active interventions result in challenges, the feedback of which alters such an internal model. I have made similar arguments, emphasizing the importance of the self and the world model established within the vestibulocerebellar system (see chapter 3). A second explanation for the role of action is that many brain learning subsystems are in essence spatial maps of the world or of the self (for example, consider Merzenich et al.'s (1984) research reported in chapter 3), and these networks of neurons essentially replicate in their functional qualities aspects of the self or the world. It must be concluded, however, that we still do not fully understand how spontaneous action results in increased adaptive learning.

REFERENCES

Adams, R. B. & Victor, M. (1985), Normal development and deviations of the nervous system. In: *Principles of Neurology*, (3rd ed.,) ed. R. B. Adams & M. Victor. New York: McGraw Hill, pp. 419–448.

Agnati, L. F. & Fuxe, K. (1984), New concepts on the structure of neuronal networks: The miniaturization and hierarchical organization of the central nervous system. *Biosci. Rep.* 4:2, 93–98.

Agranoff, B. W. (1978), Biochemical events mediating the formation of short-term memory. In: *Neurological Basis of Learning and Memory*, ed. Y. Tsukuda & B. W. Agranoff. Chichester: Wiley, pp. 135–147.

Aguayo, A. J. (1985), Capacity for renewed axonal growth in the mammalian central nervous system. In: Central Nervous System Plasticity and Repair, ed. A. Bignami, F. E. Bolm, C. L. Bolis & A. Adeloye, New York: Raven Press, pp. 31–40.

Albus, J. S. (1981), *Brain, Behavior, and Robotics*. Peterborough, NH: BYTE Books.

Alexander, T. (1985), Artificial intelligence. *Popular Computing*, pp. 68–69, 142–145.

Alkon, D. L. (1985), Calcium-mediated reduction of ionic currents: A biophysical memory trace. *Science*, 30:1037–1045.

Almansi, R. J. (1983), On the persistence of very early memory traces in psychoanalysis, myth and religion. *J. Amer. Psychoanal. Assn.*, 31:391–421.

Altman, L. L. (1975), *The Dream and Psychoanalysis*, rev. ed. New York: International Universities Press.

Anderson, J. A. & Rosenfeld, E., ed. (1989), *Neurocomputing: Foundations of Research*. Cambridge, MA: MIT Press.

Andreasen, N., ed. (1986), *Can Schizophrenia Be Localized in the Brain?* Washington, DC: American Psychiatric Press.

_______ (1989), Magnetic resonance imaging. In: *Brain Imaging*, ed. N. Andreasen.

Washington DC: American Psychiatric Press, pp. 67–122.

Annett, M. (1985), *Left, Right Hand and Brain: The Right Shift Theory.* Hillsdale, NJ: Lawrence Erlbaum Associates.

Antrobus, J., Ehrlichman, H., Weiner, M. & Wollman, M. (1982), The REM report and the EEG: Cognitive processes associated with cerebral hemispheres. In: *Sleep,* ed. W. P. Koella. Basel: Karger, pp. 49–51.

Anzieu, D. (1975), El cuerpo y el código en los cuentos de J. L. Borges. *Rev. de. Occidente,* 143:4.

Arbib, M. A., Kilmer, W. L. & Spinelli, D. N. (1976), Neural models of memory. In: *Neural Mechanisms of Learning and Memory,* ed. M. R. Rosenzweig & E. L. Bennett. Cambridge, MA: MIT Press, pp. 109–132.

Arieti, S. (1974), The rise of creativity: From primary to tertiary process. *Contemp. Psychoanal.,* 1:51–68.

Arkin, A. M. (1978), Editor's commentary on Chapter 16. In: *The Mind in Sleep,* ed. A. M. Arkin, J. Antrobus & S. J. Ellman. Hillsdale, NJ: Lawrence Erlbaum Associates, pp. 542-546.

Arlow, J. A. (1969), Unconscious fantasy and disturbances of conscious experience. *Psychoanal. Quart.,* 38:1–27.

Aschoff, J. (1981a), Short-term rhythms in activity. In: *Handbook of Behavioral Neurobiology, Vol. 4: Biological Rhythms.* New York: Plenum Press, pp. 491–498.

_______ ed. (1981b), *Handbook of Behavioral Neurobiology, Vol. 4: Biological Rhythms.* New York: Plenum Press.

Aserinsky, E. & Kleitman, N. (1953), Regularly occuring periods of eye motility and concomitant phenomena during sleep. *Science,* 118:273–74.

Bach-y-Rita, P. (1981), Brain plasticity as a basis of the development of rehabilitation procedures for hemiplegia. *Scand. J. Rehab. Med.,* 13:73–83.

Bakan, P. (1978), Dreaming, REM sleep and the right hemisphere: A theoretical integration. *J. Altered States of Consciousness,* 3:285–307.

Bard, P. (1934), Emotion 1: The neurohumoral basis of emotional reactions. In: *Handbook of General Experimental Psychology,* ed. C. A. Murchison. Worcester, MA: Clark University Press, pp. 264–311.

Bartus, R. T., Dean, R. L. Sherman, K. A., Friedman, D. & Beer, B. (1981), Profound effects of combining choline and piracetam on memory enhancement and cholinergic function in aged rats. *Neurobiology of Aging, 2:105–111.*

Basch, M. F. (1975), Perception, consciousness, and Freud's "Project." *The Annual of Psychoanalysis,* 3:3–20. New York: International Universities Press.

_______ (1976a), Psychoanalysis and communications science. *The Annual of Psychoanalysis,* 4:385–421. New York: International Universities Press.

_______ (1976b), Psychoanalytic interpretation and cognitive transformation. *Internat. J. Psycho-Anal.,* 62:151–175.

_______ (1979), Mind, self, and dreamers. In: *Sexual Excitement,* ed. R. J. Stoller. New York: Pantheon, pp. 224–231.

_______ (1983), The perception of reality and the disavowal of meaning. *The Annual of Psychoanalysis,* 11:125–154. New York: International Universities Press.

_______ (1985), Development and defense in psychotherapeutic intervention in adolescence. Presented at meeting of the American Society of Adolescence,

Dallas, Texas, May 17 (Tapes # 1, 204-1A, B, C available through Infomedix, Garden City, CA 92643).

Bateson, G. (1955), A theory of play and fantasy. In: *Semlotics: An Introductory Anthology*. ed. R. E. Innis. Bloomington: Indiana University Press, pp. 131–144.

Bellugi, U., Klima, E. S. & Poizner, A. (1988), Sign language and the brain. In: *Language, Communication, and the Brain*, ed. F. Plum. New York: Raven Press, pp. 39–56.

Benedek, T. (1937/1938), Adaptation to reality in early infancy. *Psychoanal. Quart.*, 7:200–215.

Benson, D. F. (1986), Aphasia and the lateralization of language. *Cortex*, 22:71–86.

______ & Stuss, D. T. (1989), Theories of frontal lobe function. In: *Neurology and Psychiatry*, ed. J. Meller. Basel: Karger, pp. 266–283.

Benvenuto, B. & Kennedy, R. (1986), *The Works of Jacques Lacan*. New York: St. Martin's Press.

Berger, L. (1967), Function of dreams. *J. Abnorm. Psych.* Monogr. 5:1–28.

Berger, R. J. (1970), REM sleep and mechanisms of oculomotor control. In: *Sleep and Dreaming*, ed. E. Hartmann. Boston: Little, Brown, pp. 277–294.

Bergmann, M. S. & Jucovy, M. E. (1982), *Generations of the Holocaust*. New York: Basic Books.

Berman, K. F., Zec, R. G. & Weinberger, D. R. (1986), Physiologic dysfunction of dorsolateral prefrontal cortex in schizophrenia II: Role of neuroleptic treatment, attention, and mental effort. *Arch. Gen. Psychiat.*, 43:126–135.

Bertini, M. (1982), Individual differences in the information-processing modes in sleep and waking states. In: *Sleep 1982*, ed. W. P. Koella. Basel: Karger, pp. 57–62.

Bignami, A., Bolm, F. A., Bolis, C. L. & Adeloye, A., eds. (1985), *Central Nervous System Plasticity and Repair*. New York: Raven Press.

Black, I. B. (1984), Intraneuronal mutability: Implications for memory mechanisms. *Brain Behav. Evol.*, 24:35–46.

Black, M. M. (1985), Ontogenetic determinants of regenerative vigor. In: *Central Nervous System Plasticity and Repair*, ed. A. Bignami, F. A. Bolm, C. L. Bolis & A. Adeloye. New York: Raven Press, pp. 25–30.

______ Adler, J. E., Dreyfus, C. F., Jonakait, G. M., Katz. D. M., LaGamma, E. F. & Markey, K. M. (1984), Neurotransmitter plasticity at the molecular level. *Science*, 225:1266–1270.

Bloom, F. E. (1985), CNS plasticity: A survey of opportunities. In: *Central Nervous System Plasticity and Repair*, ed. A. Bignami, F. E. Bolm, C. L. Bolis & A. Adeloye. New York: Raven Press, pp. 3–11.

Blum, H. (1977), The prototype of preoedipal reconstruction. *J. Amer. Psychoanal. Assn.*, 25:757–781.

Boden, M. A. (1988), *Computer Models of the Mind*. Cambridge: Cambridge University Press.

Borberly, A. A. (1986), Endogenous sleep substances and sleep regulation. *J. Neural Transm.* (suppl.) 21:243–254.

Boschan, P. J. (1986), Attention, interpretation, identity, and narcissism. *Internat. J. Psycho-Anal.*, 70:255–263.

Bower, G. H. & Morrow, D. G. (1990), Mental models in narrative comprehension.

Science, 5;247(4938):48–8.

Breuer, J. & Freud, S. (1893–1895), Studies on hysteria. *Standard Edition,* 2:1–305. London: Hogarth Press, 1955.

Brockes, J. P. (1984), Mitogenic growth factors and nerve dependence of limb regeneration. *Science,* 225:1280–1287.

Broughton, R. J. (1975), Biorhythmic variations in consciousness and psychological function. *Candian Psych. Ref.,* 16:217–239.

Brown, J. (1969), Hemispheric specialization and the corpus callosum. In: *Present Concepts in Internal Medicine,* ed. C. Gunderson. San Francisco: Letterman Hospital, pp. 77–86.

_______ (1972), *Aphasia, Apraxia and Agnosia.* Springfield, IL: Charles C. Thomas.

_______ (1975), The neural organization of language: Aphasia and neuropshychiatry. In: *American Handbook of Psychiatry,* 2nd ed., ed. M. F. Reiser. 4:244-298, New York: Basic Books.

_______ (1974), Language, cognition and the thalamus. *Confin. Neurol.,* 36:33–60.

Brozoski, T., Brown, R. M., Rosvold, H. E. & Goldman, P. S. (1979), Cognitive deficit caused by depletion of dopamine in prefrontal cortex of rhesus monkey. *Science,* 205:929–931.

Buchsbaum, M. S., DeLisi, L. E. & Holcomb, H. H. (1983), Anterior-posterior gradients in cerebral glucose use in schizophrenia and affective disorders. *Arch. Gen. Psychiat.* 41:1159–1166.

Bühler, K. (1982), The key principle: The sign character of language. In: *Semiotics,* ed. R. E. Innis. Bloomington: Indiana University Press, 1985, pp. 66–86.

Burghardt, G. M. (1988), Precocity, play, and the ectoderm-endoderm transition: Profound reorganization or superficial adaptation? In: *Handbook of Behavioral Neurobiology, Vol. 9: Developmental Psychobiology and Behavioral Ecology,* ed. E. M. Blass. New York: Plenum Press, pp. 107–148.

Call, J. D. (1980), Some prelinguistic aspects of language development. *J. Amer. Psychoanal. Assn.,* 28:259–280.

Callaway, E. & Dembo, D. (1958), Narrowed attention. *Amer. Archs. Neur. Psychiat.,* 79:74–90.

Cannon, W. B. (1927), Some basic mechanisms of the translation of bodily needs into behavior. In: *CIBA Symposium on the Neurological Basis of Behavior,* ed. G. E. W. Wolstenholme & C. M. O'Connor. Boston: Little, Brown, pp. 187–200.

Carrol, R. L. (1988), *Vertebrate Paleontology and Evolution.* New York: W. H. Freeman.

Charniak, E. & McDermott, D. ed. (1985), *Introduction to Artifical Intelligence,* Reading, MA: Addison-Wesley.

Chomsky, N. (1972), *Language and Mind.* New York: Harcourt.

_______ (1975), *Reflections on Language.* New York: Pantheon.

Chugani, H. T. & Phelps, M. E.(1986), Maturational changes in cerebral function in infants determined by FDG positron emission tomography. *Science,* 231:840–843.

Clancey. W. J. & Shortliffe, E. H., eds. (1984), *Readings in Medical Artificial Intelligence.* Reading, MA: Addison-Wesley.

Cleghorn, R. A. (1980), Endorphine-morphine-like peptides of brain, *Canad J.*

Psychiat. 25:182–186.

Cohen, P. R. & Feigenbaum, E. A., eds. (1982), *The Handbook of Artificial Intelligence, Vol.3.* Stanford, CA: Harris Tech Press.

Collins, N. (in press). Psychotherapy and minimum language skills: Mutual exclusion? In: *At the Crossroads: A Celebration of Diversity,* ed. D. Watson. Little Rock, AR: American Deafness and Rehabilitation Assn.

Cooper, I. S., ed. (1978), *Cerebellar Stimulation in Man.* New York: Raven Press.

______ Ricklan, M., Amin, I. & Cullinen, T. (1978), A long-term follow-up study of cerebellar stimulation for the control of epilepsy. In: *Cerebellar Stimulation in Man,* ed. I. S. Cooper. New York: Raven Press, pp. 19–38.

Corhalles, M. (1978), On the biological basis for laterality I: Evidence for a left-right maturational gradient. *Behav. Brain Sci.* 1:261–269.

Cotman, C. W. & Nieto-Sampedro, M. (1984), Cell biology of synaptic plasticity. *Science,* 225:1287–1294.

Cowan, W. M. (1979), The development of the brain. *Sci. Amer,* 241:112–133.

______ Fawcett, J. W., O'Leary, D. D. M. & Stanfield, B.B. (1984), Regressive events in neurogenesis. *Science,* 225:1258–1265.

Crow, T. J. (1985), The two-syndrome concept: Origins and current status, *Schizophren. Bull.* 11:471–5.

______ (1986a), Left brain, retrotransponsons, and schizophrenia. *Br. Med. J.,* 293:3–4.

______ (1986b), Secular changes in affective disorders and variations in the psychosis gene. *Arch. Gen. Psychiat.,* 43:1012–1014.

______ (1986c), The continuum of psychosis and its implications for the structure of the gene. *Brit. J. Psychiat.,* 149:419–29.

Darwin, C. (1872), *The Expression of the Emotions.* New York: Philosophic Library, 1955.

Decety, J. & Ingvar, D. (1988), Symposium on the principles and operation of the brain. Presented at the Pontifical Academy, Rome, Italy.

Demos, E. V. (1985), The revolution in infancy research: Its implication for the modification of developmental theory. Presented at meeting of the American Society of Adolescence, Dallas, May 17 (Tapes # 1, 204–1 A, B, C available through Infomedix, Garden City, CA 92643.)

Desmedt, J. E. (1979), *Progress in Clinical Neurophysiology, Vol. 6: Cognitive Components in Cerebral Event-Related Potentials and Selective Attention.* Basel: Karger.

DeRobertis, E. M., Oliver, G. & Wright, C. V. E. (1990), Homeobox genes and the vertebrate body plan. *Scientific Amer.,* 263:46–53.

DeVoogd, T. & Nottebohm, F. (1981), Gonadal hormones induce dendritic growth in the adult avian brain. *Science,* 24:202–204.

Dewan, E. M. (1970), The programming hypothesis for REM sleep. In: *Sleep and Dreaming,* ed. E. Hartmann. Boston: Little, Brown, pp. 295–307.

Donchin, E., Ritter, W. & McCallum, W. C. (1978), Cognitive psychophysiology: The endogenous components of ERP. In: *Event-Related Brain Potentials in Man,* ed. E. Callaway, P. Tueting & S. H. Koslow. New York: Academic Press.

Drachman, D. A. (1977), Memory and cognitive functions in man: Does the cholinergic system have a specific role? *Neurology,* 27:787–790.

Dushenko, T. W. & Sterman, M. B. (1984), Hemisphere-specific deficits on cognitive/perceptual tasks following REM sleep deprivation. *Int. J. Neurosci.*, 25:25–45.
Eccles, J. (1978), *The Human Psyche.* New York: Springer.
_______ (1979), Introductory remarks. In: *Cerebro-Cerebellar Interactions,* ed. J. Massion & K. Sasaki. Amsterdam: Elsevier/North Holland Biomedical Press, pp. 1–18.
_______ (1982), The future studies on the cerebellum. In: *The Cerebellum,* ed. S. L. Palay & V. Chan-Palay. Berlin: Springer, pp. 607–620.
_______ Szentagothai, J. & Itoh, M. (1967), *The Neuronal Machine.* New York: Springer.
Edelman, G. M. (1983), Cell adhesion molecules. *Science,* 219:450–457.
_______ & Chuong, C. (1982), Embryonic to adult conversion of neural cell adhesion molecules in normal and staggerer mice, *Proc. National Acad. Science,* 79:7036–7040.
Ehrlichman, H., Antrobus, J. S. and Weiner, M. S. (1985), EEG asymmetry and sleep mentation during REM and NREM. *Brain Cogn.* 4:477–485.
Einstein, A. (1926), The cause of the formation of meandering in the course of rivers. In: *Einstein: A Centenary Volume,* ed. A. P. French. Cambridge, MA: Harvard University Press, pp. 298–301.
Embde, R. (1983), The prerepresentational self and its affective core. *The Psychoanalytic Study of the Child,* 38:165–192. New Haven, CT: Yale University Press.
Feldman, C. F. & Toulmin, S. Logic and the theory of mind. Presented at the Nebraska Symposium on Motivation, 1974–1975.
Feldman, S. S. (1959), *Mannerisms of Speech and Gesture in Everyday Life.* New York: International Universities Press.
Ferenczi, S. (1926), An attempted explanation of some hysterical stigmata. In: *Further Contributions to the Theory and Technique of Psychoanalysis.* London: Hogarth Press, pp. 110–117.
Fine, H. J., Pollio, H. R. & Simpkinson, C. H. (1972), Figurative language, metaphor and psychotherapy. *Psychother. Res. Pract.,* 10:87–91.
Fisher, C., Byrne, J., Edwards, A. & Kahn, E. (1970), A psychophysiological study of nightmares. *J. Amer. Psychoanal. Assn.,* 18:747–782.
Flor-Henry, P. (1983), *Cerebral Basis of Psychopathology.* Littleton, MA: Wright-PSG.
Fónagy, E. (1983), *La Vive Voix: Essais de Psycho-Phonêtique.* Paris: Payot.
Forest, D. V. (1973), On one's own onymy. *Psychiat.* 36:266–290.
Foulkes, D. (1983), General discussion: Dream psychology. In: *Sleep Disorders,* ed. M. Chase & E. D. Weitzman. New York: Spectrum, pp. 401–413.
Fraser, J. G. (1890), *The New Golden Bough,* ed. T. H. Gaster. New York: New American Library, 1964.
French, T. M. (1952), *The Integration of Behavior, Vol. 1.* Chicago: University of Chicago Press.
Freud, S. (1888), Preface to the translation of Bernheim's *Suggestion. Standard Edition,* 1:73–85. London: Hogarth Press, 1966.
_______ (1891a), On the interpretation of the aphasias, a critical study. *Standard Edition,* 3:240–241. London: Hogarth Press, 1962.
_______ (1891b), *On Aphasia,* trans. E. Stengel. New York: International University Press, 1953.

_______ (1895), Project for a scientific psychology. *Standard Edition,* 1:295–397. London, Hogarth Press, 1966.

_______ (1897), Letter to Fleiss, November 14. *Standard Edition.* 1:268–271. London: Hogarth Press, 1966.

_______ (1900), The interpretation of dreams. *Standard Edition,* 4 & 5. London: Hogarth Press, 1953.

_______ (1901), The psychopathology of everyday life. *Standard Edition,* 6. London: Hogarth Press, 1960.

_______ (1905), Jokes and their relation to the unconscious. *Standard Edition,* 8. London: Hogarth Press, 1960.

_______ (1911), Formulations on the two principles of mental functioning. *Standard Edition,* 12:218–226. London: Hogarth Press, 1958.

_______ (1914a), On the history of the psycho-analytic movement. *Standard Edition,* 14:7–66. London: Hogarth Press, 1957.

_______ (1914b), On narcissism: An introduction. *Standard Edition,* 14:73–102. London: Hogarth Press, 1957.

_______ (1914c), The Moses of Michelangelo. *Standard Edition,* 13:211–238. London: Hogarth Press, 1964.

_______ (1915), The unconscious. *Standard Edition,* 14:166–204. London: Hogarth Press. 1957.

_______ (1916/1917), Introductory lectures on psychoanalysis. *Standard Edition,* 15 & 16. London: Hogarth Press, 1963.

_______ (1920), Beyond the pleasure principle. *Standard Edition,* 18:7–64. London: Hogarth Press. 1955.

_______ (1923), The ego and the id. *Standard Edition,* 19:12–66. London: Hogarth Press, 1961.

Friberg, L. & Roland, P. E. (1987), Functional activation and inhibition of regional cerebral blood flow and metabolism. In: *Basic Mechanisms of Headache,* ed. J. Olesen & L. Edvinsson. Amsterdam: Elsevier, pp. 2–30.

Frick, R. B. (1982), The ego and the vestibulo-cerebellar system: Some theoretical perspectives. *Psychoanal. Quart.,* 51:95–122.

Friedman, S. & Fisher, C. (1967), On the presence of a rhythmic, diurnal, oral instinctual drive cycle in man: A preliminary report. *J. Amer. Psychoanal. Assn.,* 15:17–343.

Fromkin, V. A. (1988), The state of brain/language research. In: *Language, Communication and the Brain,* ed. F. Plum. New York: Raven Press, pp. 1–18.

Fromm, E. (1947), *Man for Himself.* New York: Rinehart.

Fukuda, M., Niwa, S., Hiramatsu, K., Hayashida, S., Saitoh, O., Kageyama, T., Nakagome, K., Iwanami, A., Sasaki, T. & Itoh, K. (1989), Psychological intervention can partly alter P300 amplitude abnormalities in schizophrenics. Unpublished manuscript.

Funt, D. (1973), The question of the subject: Lacan and psychoanalysis. *Psychoanal. Rev.,* 60:393–405.

Fuster, J. M. (1980), *The Prefrontal Cortex.* New York: Raven Press.

Gabel, S. (1985), Sleep research and clinical reported dreams: Can they be integrated? *J. Anal. Psychol.,* 30:185–205.

Galaburda, A. M., LeMay, M., Kemper, T. L. & Geschwind, N. (1978), Right-left asymmetries in the brain. *Science,* 199:852–856.

Galin, D. (1974), Implications for psychiatry of left and right cerebral specialization. *Arch. Gen. Psychiat.*, 31:572–583.

_______ & Ornstein, R. (1974), Individual differences in cognitive style I: Reflective eye movements. *Neuropsychologia*, 12:367–376.

Gall, F. & Spurzheim, C, (1810–1819), *Anatomie et Physiologie du Système Nerveux en Géneral et du Cerveau en Particulier.* Paris: Schoell.

Gans, E. (1981), *The Origin of Language.* Berkeley: University of California Press.

Gasanov, U. G. (1984), Study of higher nervous activity and current neurophysiology. *Neuroscience and Behav. Psychol.*, 14:187–194.

Gastaut, H. & Broughton, R. (1965), A clinical and polygraphic study of episodic phenomena during sleep. In: *Recent Advances in Biological Psychiatry, Vol. 7*, ed. J. Wortis. New York: Plenum, pp. 197–220.

Gazzaniga, M. (1970), *The Dissected Brain.* New York: Appleton-Century Crofts.

_______ (1976), The biology of memory. In: *Neural Mechanisms of Learning and Memory*, ed. M. R. Rosenzweig & E. L. Bennett. Cambridge, MA: MIT Press, pp. 57–66.

Gazzaniga, M. (ed.) (1979), *Handbook of Behavioral Neurobiology, Vol. 2, Neuropsychology.* New York and London: Plenum Press.

_______ & Hillyard, S. A. (1971), Language and speech capacity in the right hemisphere. *Neuropsychologia*, 9:273–280.

Gedo, J. (1978), The analyst's affectivity and the management of the transference. Presented to the Chicago Psychoanalytic society.

_______ (1979), *Beyond Interpretation.* New York: International Universities Press.

_______ (1981), *Advances in Clinical Psychoanalysis.* New York: International Universities Press.

_______ (1984a), *Psychoanalysis and its Discontents.* New York: Guilford Press.

_______ (1984b), Discussion of Joseph Lichtenberg's "The Empathic Mode of Perception and Alternative Vantage Points for Psychoanalysis." In: *Empathy II*, ed. J. Lichtenberg, M. Bernstein & D. Silver. Hillsdale, N. J: The Analytic Press, pp. 137–142.

_______ (1986), *Conceptual Issues in Psychoanalysis: Essays in History and Method.* Hillsdale, NJ: The Analytic Press.

_______ (1989a), Psychoanalysis and Occam's Razor. Presented to the Chicago Psychoanalytic Society.

_______ (1989b), An epistemology of transference. *The Annual of Psychoanalysis*, 17:3–16. Hillsdale, NJ: The Analytic Press.

_______ & Goldberg, A. (1973), *Models of the Mind.* Chicago: University of Chicago Press.

Geschwind, N. (1964), The paradoxical position of Kurt Goldstein in the history of aphasia. *Cortex*, 1:214–224.

_______ (1968), Isolation of the speech area. *Neuropsychologia*, 6:327–340.

_______ (1972), Language and the brain. *Sci. Amer.*, 226:76–83.

Gibson, W. C. (1962), Pioneers in localization of function in the brain. *J. Amer. Med. Assn.*, 180:122–129.

Gilbert, P. (1975), How the cerebellum could memorize movements. *Nature*, 254:688–689.

_______ (1976), Letter in response to a letter from A. Routenberg. *Nature*, 260:80

Gill, M. M. (1967), The primary process. *Psychological Issues*, Monogr. 18/19, pp. .

258–294. New York: International Universities Press.

Gilles, F. H., Leviton, A. & Dooling, E. C., eds. (1983), *The Developing Human Brain.* Boston: John Wright-PSG.

Giora, Z. (1974), What a dream is. *Brit. J. Med. Physiol.,* 47:283–289.

Goldman, A. I. (1986), *Epistemology and Cognition.* Cambridge, MA: Harvard University Press.

Goldman-Rakic, P. S., Isseroff, A., Schwartz, M. L. & Bugbee, N. M. (1983), Neurobiology of cognitive development in non-human primates. In: *Handbook of Child Psychology,* ed. P. Mussen. *4th Edition, Vol. 2: Infancy and Developmental Psychobiology,* ed, M. M. Haith & J. J. Campos. New York: Wiley, pp. 281–344.

Goodall, J. & Hamburg, D. A. (1975), Chimpanzee behavior as a model for the behavior of early man: New evidence on possible origins of human behavior. In: *American Handbook of Psychiatry, Vol. 6,* ed. S. Arieti. New York: Basic Books, pp. 14–43.

Gould, S. J. (1977), *Ontogeny and Phylogeny.* Cambridge, MA: T. Belknap Press/ Harvard University.

Greenberg, R. (1970), Dreaming and memory. In: *Sleep and Dreaming,* ed. E. Hartmann. Boston: Little, Brown, pp. 258–267.

Greitz, T., Ingvar, D. H. & Widen. L., eds. (1985), *The Metabolism of the Human Brain Studied with Positron Emission Tomography.* New York: Raven Press.

Grinker, R. R., Sr. (1975), *Psychiatry in Bold Perspective.* New York: Behavioral Publications.

Gross, G. (1982), Regulation of the circadian sleep-wake cycle. In: *Sleep 1982,* ed. W. P. Koella. Basel: Karger, pp. 19–29.

Grotstein, J. S. (1981), Some newer developments in Kleinian theory and practice. Presented at annual meeting of the American Psychoanalytic Association, New York City.

Gunther, M. (1987), Discussion of Levin and Vuckovich's paper "Psychoanalysis and the Two Cerebral Hemispheres." Presented to the Chicago Psychoanalytic Society.

Gur, R. C. & Gur, R. E. (1980), Handedness and individual differences in hemispheric activation. In: *Neurophysiology of Left Handedness,* ed. J. Herron. New York: Academic Press, pp. 211–231.

Hadley, J. L. (1985), Attention, affect, and attachment. *Psychoanal. Contemp. Thought,* 8:529–550.

_______ (1987), Discussion of Levin and Vuckovich's paper "Psychoanalysis and the Two Cerebral Hemispheres." Presented to the Chicago Psychoanalytic Society.

Hagan, S. (1984), Neurobiology's summer place. *Outlook Magazine,* Spring: 2–5.

Hagman, R. S. (1982), *Language, Life and Human Nature.* Carbondale, IL: Linguistic Research.

Hamilton, N. G., Frick, R. B., Takahashi, T. & Hopping, M. W. (1983), Psychiatric symptoms and cerebellar pathology. *Amer. J. Psychiat.,* 140:1322–1326.

Hartmann, H. (1939), *Ego Psychology and the Problem of Adaptation,* trans. D. Rapaport. New York: International Universities Press.

_______ (1964), *Essays on Ego Psychology.* New York: International Universities Press.

_______ Kris, E. & Loewenstein, R. M. (1953), The function of theory in psychoanalysis. In: *Drives, Affects, Behavior,* ed. R. M. Loewenstein. New York: International Universities Press, pp. 13–37.

Heath, R. G. (1977), Modulation of emotion with a brain pacemaker. *J. Nerv. Ment. Dis.,* 165:300–317.

Hebb, D. O. (1955), Drives and the C. N. S. (conceptual nervous system). *Psychol. Rev.,* 62:243–254.

Heinrichs, J. & Endicott, K. (1988), Music and the mind: Modelling the brain. *Dartmouth Alumni Magazine,* 82:18–23.

Henderson, J. L. & Wheelwright, J. B. (1974), Analytic psychology. In: *American Handbook of Psychiatry, Vol. 1,* ed. S. Arieti. New York: Basic Books, pp. 809–819.

Hermann, I. (1934), Ur Wahrnehmungen, In sbesondere Augen leuchten und Lautwerden des inneren. *Int. z. Psychoanal,* 20:553–555.

Hermann, I. (1936), Clinging and going-in-search: A contrasting pair of instincts and their relation to sadism and masochism. *Psychoanal. Quart,* 45:5–36, 1976.

Hewes, G. W. (1973), Primate communication and the gestural origin of language. *Current Anthropol.,* 14:5–24.

Hill, D. (1981), Mechanisms of the mind: A psychiatrist's perspective. *Br. J. Med. Psychol.,* 54:1–13.

Hillyard, S. A. (1979), Event-related brain potentials and selective information processing in man. In: *Progress in Clinical Neurophysiology, Vol. 6,* ed. J. E. Desmedt. Basel: Karger, pp. 1–52.

Hiramatsu, K., Akimoto, M., Fukuda, M., Saitoh, O., Kameyama, T., Niwa, S. & Itoh, K. (1986), Decoupling of stimulus and response processes seems responsible for slow reaction times in schizophrenics with mild symptomatology. In: *Brain Electrical Potentials and Psychopathology,* ed. C. Shagrass, R. C. Josiassen & R. A. Roemer. Amsterdam: Elsevier, pp. 183–193.

Hobson, J. A. & McCarley, R. W. (1977), The brain as a dream state generator: An activation-synthesis hypothesis of the dream process. *Amer. J. Psychiat.,* 134:1335–1348.

Hoffman, S. & Edelman, G. M. (1983), Kinetics of homophilic binding of embryonic and adult forms of the neural cell adhesion molecule. *Proc. Natl. Acad. Sci.* USA, 80:5762–5766.

Hoit, M. (1984), Diagnostic implications of the collapse of the work-ego in adolescence. In: *Late Adolescence: Psychoanalytic Studies,* ed. D. D. Brockman. New York: International Universities Press, pp. 227–240.

Holton, G. (1979), What precisely is 'thinking'? Einstein's answer. In: *Einstein: A Centenary Volume,* ed. A. P. French. Cambridge, MA: Harvard University Press, pp. 153–166.

Ingvar, D. (1983a), Abnormalities of activity distribution in the brain of schizophrenics: A neurophysiological interpretation. In: *Perspectives in Schizophrenia Research,* ed. D. Baxter, & B. Melneshuk. New York: Raven Press, pp. 107–130.

_______ (1983b), Serial aspects of language and speech related to prefrontal cortical activity: A selective review. *Human Neurobiology* 2:177–189.

_______ & Franzen, G. (1974), Abnormalities of cerebral blood flow distribution in patients with chronic schizophrenia. *Acta Psychiatr. Scand.,* 15:425–462.

Innis, R. E. (1985), *Semiotics.* Bloomington: Indiana University Press.

Isay, R. A. (1977), Ambiguity in speech. *J. Amer. Psychoanal. Assn.,* 25:427–452.

Itoh, M. (1970), Neurophsyiological aspects of the cerebellar motor control system. *Internat. J. Neurol.,* 7:162–176.

_______ (1976), Cerebellar learning control of vestibular-ocular mechanisms. In: *Machanisms in Transmission of Signals for Conscious Behavior,* ed. T. Desiraju. Amsterdam: Elsevier, pp. 1–22.

_______ (1981), *Blueprints of the Brain.* Tokyo: Shizen.

_______ (1984a), *The Cerebellum and Neural Control.* New York: Raven Press.

_______ (1984b), Cerebellar plasticity and motor learning. *Exp. Brain Res.,* 9:165–169.

_______ (1984c), The modifiable neuronal network of the cerebellum. *Jpn. J. Physiol.,* 34:781–792.

_______ (1985a), Memory system in the cerebellum. In: *Perspectives on Neuroscience: From Molecule to Mind,* ed. Y. Tsukada. Tokyo: University of Tokyo Press, pp. 214–235.

_______ (1985b), Synaptic plasticity in the cerebellar cortex that may underlie the vestibulo-ocular adaptation. In: *Adaptive Mechanisms in Gaze Control,* ed. A. Berthoz & G. M. Jones. Amsterdam: Elsevier, pp. 213–221.

_______ (1988), Neural control as a major aspect of higher-order brain function. Presented to the Pontifical Academy symposium on Principles of Design and Operation of the Brain, Rome, Italy.

Jackson, J. H. (1958), *Selected Writings,* ed. J. Taylor. New York: Basic Books.

Jakobson, R. (1960), Linguistics and poetics. In: *Semiotics,* ed. R. E. Innis. Bloomington: Indiana University Press, 1985, pp. 142–175.

_______ (1987), Two aspects of language and two types of aphasic disturbances. In: *Language in Literature,* ed. K. Pomorska & S. Rudy. Cambridge, MA: Belknap Press, Harvard University, pp. 95–114.

James, W. (1890), *The Principles of Psychology, Vol. 1.* New York: Dover Press.

Jeannerod, M. (1985), *The Brain Machine,* trans. D. Urion. Cambridge, MA: Harvard University Press.

Joseph, R. (1982), The neuropsychology of development: hemispheric laterality, limbic language, and the origin of thought, *J. Clin. Psychol.* [special monograph supplement], 38:4–33.

Jouvet, M. (1982), Hypnogenic indolamine-dependent factors and paradoxical sleep rebound. In: *Sleep 1982,* ed. W. P. Koella. Basel: Karger, pp. 21–28.

Kahn, E., Fisher, C. & Edwards, A. (1978), Night terrors and anxiety dreams. In: *The Mind in Sleep,* ed. A. M. Arkin, J. Antrobus & S. J. Ellman. Hillsdale, NJ: Lawrence Erlbaum Associates, pp. 533–542.

Kandel, E. (1983), From metapsychology to molecular biology: Exploration into the nature of anxiety. *Amer. J. Psychiat.,* 140:1277–1293.

Kaas, J. H., Krubitzer, L. H., Chino, Y. M., Landstrom, A. L., Polley, E. H. & Blair, N. (1990), Reorganization of retinotopic cortical maps in adult mammals after lesions of the retina. *Science,* 248:229–231.

Kent, E. (1981), *The Brains of Men and Machines.* Peterborough, NH: BYTE.

Kety, S. (1982), The evolution of concepts of memory. In: *The Neural Basis of Behavior,* ed. A. L. Beckman. Jamaica, NY: Spectrum, pp. 95–101.

Kinsbourne, M. (1972), Eye and head turning indicates cerebral lateralization. *Science,* 176:539–541.

Kinsbourne, M. (1980), A model for the ontogeny of cerebral organization in non-right handers, In: *Neurophysiology of Left-Handedness*, ed. J. Herron, New York: Academic Press, pp. 177–185.

Kitney, R. I., Miall, R. C., Stein, J. F. & Riddell, P. M. (1984), Time series analysis of neuronal signals recorded in the cerebellum of trained monkeys. *J. Theor. Biol.*, 107:376–385.

Klein, G. (1966), The several grades of memory. In: *Psychoanalysis—A General Psychology: Essays in Honor of Heinz H. Hartmann*, ed. R. M. Loewenstein, L. M. Newman, M. Schurr & A. J. Solnit. New York: International Universities Press, pp. 377–389.

Klein, R. & Armitage, R. (1979), Rhythms in human performance: 1 and 1/2 hour oscillations in cognitive style. *Science*, 204:1326–1328.

Kleitman, N. (1963), *Sleep and Wakefulness*. Chicago: University of Chicago Press.

Koh, S. D. & Kayton, L. (1974), Memorization of 'unrelated' word strings by young nonpsychotic schizophrenics. *J. Abnorm. Psych.*, 83:14–22.

Kohonen, T. (1983), Representation of information in spatial maps which are produced by self organization. In: *Synergetics of the Brain*, ed. E. Basar, H. Flohr, H. Haken & A. J. Mandell. Berlin: Springer, pp. 264–273.

Kohut, H. (1959), Introspection, empathy and psychoanalysis. *J. Amer. Psychoanal. Assn.* 7:459–483.

_______ (1966), Forms and transformations of narcissism. *J. Amer. Psychoanal. Assn.* 14:243–272.

_______ (1971), *The Analysis of the Self*. New York: International Universities Press.

_______ (1977), *The Restoration of the Self*. New York: International Universities Press.

Kolata, G. (1984), Fernando Nottebohm: New neurons form in adulthood, *Science*, 224:22, 1325–1326.

Kolb, B. & Whinshaw, I. Q. (1980), The frontal lobes. In: *Fundamentals of Human Neuropsychology*, ed. B. Kolb, C. Q. Whinshaw, San Francisco: W. H. Freeman, pp. 277–307.

Kreiger, D. T., Brownstein, M. J., & Martin, J. B. ed. (1983), *Brain Peptides*. New York: Wiley.

Kripke, D. F. (1982), Ultradian rhythms in behavior and physiology, In: *Rhythmic Aspects of Behavior*, ed. F. M. Brown & R. C. Graeber. New York: Raven Press, pp. 313–343.

_______ Mullaney, D. J. & Fleck, P. A. (1984), Ultradian rhythms during sustained performance. Presented to the Symposium on Ultradian Rhythms, Munich, West Germany.

Kris, E. (1952), *Psychoanalytic Explorations in Art*. New York: International Universities Press.

Lange, C. G. & James, W. (1922), *The Emotions*. New York: Haffner, 1967.

Langer, S. K. (1942), *Philosophy in a New Key*. Cambridge, MA: Harvard University Press.

_______ (1967), *Mind: An Essay on Human Feeling, Vol. 1*. Baltimore, MD: Johns Hopkins University Press.

Lashley K. S. (1937), Functional determinants in cerebral localization. *Arch. Neurol. Psychiat.*, 38:371–387.

_______ (1938), The thalamus and emotion. *Psychol. Rev.*, 45:42–61.

_______ (1950). *Physiological Mechanisms in Animal Behavior.* New York: Academic Press.

_______ (1951), The problem of serial order in behavior. In: *Cerebral Mechanisms in Behavior,* ed. L. Jeffress. New York: Wiley, pp. 112–136.

Lassen, N. (1987), Cerebral blood flow measured by xenon-113. *Nucl. Med. Commun., 8:535–548.*

_______ Ingvar, D. H. & Skinhoj, E. (1978), Brain function and blood flow. *Sci. Amer.* 239:62–71.

Lavie, P. & Tzischinsky, O. (1986), Cognitive asymmetry and dreaming: Lack of relationship. *Amer. J. Psychol.,* 98:353–361.

Leakey, R. E. (1981), *The Making of Mankind. New York: Dutton.*

Leavy, S. A. (1973), Psychoanalytic interpretation. *The Psychoanalytic Study of the Child,* 28:305–330. New Haven: Yale University Press.

_______ (1983), Speaking in tongues: Some linguistic approaches to psychoanalysis. *Psychoanal. Quart., 52:34–55.*

_______ (1984), Review of *Lacan and Language: A Reader's Guide to Écrit* by J. P. Muller & W. J. Richardon. *Psychoanal. Quart.,* 53:582–585.

LeDoux, J. E., Wilson, D. H. & Gazzaniga, M. S. (1977), A divided mind: Observations on the conscious properties of the separate hemispheres. *Ann. Neurol.,* 2:417–421.

Lehtonen, J. (1980), The relationship between neurophysiology and psychoanalysis in the light of dream research. *Perspect. Biol. Med.,* 23:415–423.

Leider, R. J. (1984), Report of an American Psychoanalytic Association panel on: The neutrality of the analyst in the analytic situation. *J. Amer. Psychoanal. Assn.,* 32:573–586.

Levey, M. (1984/1985), The concept of structure in psychoanalysis. *The Annual of Psychoanalysis, 12/13:* 137–154. New York: International Universities Press.

Levin, F. M. (1977), How sign language is used in the service of resistance: Approaches to psychoanalytic psychotherapy with deaf adults. Colloquium on Deafness at annual meeting of the American Psychoanalytic Association, New York City.

_______ (1980), Psychoanalytically-oriented psychotherapy with the deaf. In: *Mental Health Needs of Deaf Adults and Children,* ed. L. Stein, G. Mindel & T. Jabley. New York: Grune & Stratton, pp. 113–132.

_______ (1985), The need for a psychoanalytic learning theory. Presented at meeting of the American Society of Adolescence, Dallas, May 17 (Tapes # 1, 204–1A, B, C, available through Infomedix, Garden City, CA 92643).

_______ (1988), Recent advances in understanding mentation and affect. Presented to the 44th Annual Regional Midwest Conference of the Chicago Medical Society.

_______ (1989), Discussion of J. Gedo's paper "Psychoanalytic Theory and Occam's Razor." Presented to the Chicago Psychoanalytic Society.

_______ (in press-a), Sadism and masochism in neurosis: Report of panel of the meeting of the American Psychoanalytic Association. *J. Amer. Psychoanal. Assn.*

_______ (in press-b) Psychoanalysis and the brain. In: *Psychoanalysis: The Major Concepts,* ed. B. E. Moore & B. Fine. New Haven: Yale University Press.

_______ (1990) Psychological development and the changing organization of the *brain. The Annual of Psychoanalysis,* 18:45–61. Hillsdale, NJ: The Analytic Press.

Levy, J. (1974), Psychobiological implications of bilateral symmetry. In: *Hemispheric Function in the Human Brain,* ed. S. Dimond & G. Beaumont. New York: Halsted Press, pp. 121–183.

_______ (1977), The mammalian brain and adaptive advantage of cerebral asymmetry. *Ann. NY Acad. Sci.,* 299:264–272.

_______ Trevarthen, C. (1977), Perceptual, semantic and phonetic aspects of elementary language processes in split brain patients. *Brain,* 100:105–118.

_______ (1969), The train ride: A study of one of Freud's figures of speech. *Psychoanal. Quart.,* 39:71–89.

Lewin, B. (1970), Metaphor, mind and mankind. *Psychoanal. Quart.* 40:6–39.

Lichtenberg, J. (1983), *Psychoanalysis and Infant Research.* Hillsdale, NJ: The Analytic Press.

_______ (1988), A theory of motivational-functional systems as psychic structures. *J. Amer. Psychoanal. Assn.,* 36:57–72.

_______ (1989), *Psychoanalysis and Motivation.* Hillsdale. NJ: The Analytic Press.

Lieberman, P. (1984), *The Biology and Evolution of Language.* Cambridge, MA: Harvard University Press.

Lilleskov, R. K. (1977), Report of a panel of the American Psychoanalytic Association on: Nonverbal aspects of child and adult psychoanalysis. *J. Amer. Psychoanal. Assn.,* 25:695–706.

Litowitz, B. E. (1975), Language: Waking and sleeping. In: *Psychoanalysis and Contemporary Science, Vol. 4,* ed. D. P. Spence. New York: International Universities Press, pp. 291–328.

_______ & Litowitz, N. S. (1977), The influence of linguistic theory and psychoanalysis: A critical, historical survey. Internat. Rev. Psycho-Anal., 4:419–448.

Livingston, R. B. (1976), A casual glimpse of evolution and development relating to the limbic system. In: *Limbic Mechanisms,* ed. K. E. Livingston & O. Hurnykiewicz. New York: Plenum Press, pp. 17–22.

Locke, S. (1977), Brain and Behavior. In: *Psychopathology and Brain Dysfunction,* ed. C. Shagassi, S. Gershon & A. J. Friedhoff. New York: Raven Press, pp. 69–76.

Lowenthal, F. (1982), Can apes tell us what language is? In: *Language and Language Acquisition,* ed. F. Lowenthal, F. Vandamme & J. Cordier. New York: Plenum Press, pp. 343–358.

Lusek, C. G. & Kaas, J. H. (1986), Interhemispheric connections of cortical sensory and motor representations in primates. In:*Two hemispheres-One Brain: Functions of the Corpus Callosum* ed. F. Llepore, N. Ptito & H. H. Jasper. New York: A. R. Liss, pp. 85–102.

Lynch, G & Baudry, M. (1984), The biochemistry of memory: A new and specific hypothesis. *Science,* 224:1057–1063.

MacLean, P. D. (1960), Psychosomatics. In: *Neurophysiology, Vol. 3,* ed. J. Field. Washington, DC: American Physiological Society, pp. 1723–1744.

_______ (1962), New findings relevant to the evolution of psychosexual functions of the brain. *J. Nerv. Ment. Dis.,* 135:289–301.

_______ (1985), Stepwise brain evolution with respect to vocalization and speech. Presented at meeting of the American Society to Adolescence, Dallas, May 17 (Tapes # 1, 204–1A, B, C available through Infomedix, Garden City, CA, 92643.

Makara, G. B., Palkovits, M. & Szentagothai, J. (1980), The endocrine hypothalamus and the hormonal response to stress. In:*Selye's Guide to Stress Research, Vol. 1,* ed. H. Selye. New York: Van Nostrand Reinhold, pp. 280–337.

Manning, A. (1976), Animal learning: Ethological approaches. In: *Neural Mechanisms of Learning and Memory,* ed. M. A. Rosenzweig & E. L. Bennett. Cambridge, MA: MIT Press, pp. 147–158.

Marr, D. (1969), A theory of cerebellar cortex. *J. Physiol.,* 202:437–470.

______ (1978), Representing visual information. In:*Theoretical Approaches to Neurobiology,* ed. W. E. Reichhardt & T. Poggio. Cambridge, MA: MIT Press, pp. 151–166.

______ (1984), Vision: A theory of cerebellar cortex. *J. Physiol.* (London), 202:437–470.

Massion, J & Sasaki, K., eds. (1979), *Cerebro-Cerebellar Interactions.* Amsterdam: Elsevier/North Holland Biomedical Press.

Mathew, R. J., Duncan, G. C., Weinman, M. L., & Barr, D. L. (1982), Regional cerebral blood flow in schizophrenia. *Arch. Gen. Psychiat.,* 39:1121–1124.

McClintock B. (1984), The significance of response of the genome to challenge. *Science,* 226:792–800.

McConnel, J. V. (1962), Memory transfer through cannibalism in planarians. *J. Neuropsych.,* 3:542–548.

McGlone, J. (1978), Sex differences in functional brain asymmetry. *Cortex,* 14:122–128.

McLaughlin, J. (1978), New considerations of primary and secondary process thinking in light of cerebral hemisphere lateralization studies. *PANY Bull.,* 17:2–7.

Meissner, W. (1966), Hippocampal functions in learning. *J. Psychiatr. Res.,* 4:235.

Merzenich, M. M., Randall, J. N., Stryker, M. P., Cynader, M. S., Schoppmann, A. & Zook, J. M. (1984), Somatosensory cortical map changes following digit amputation in adult monkeys. *J. Comp. Neurol.,* 224:591–605.

Mesulam, M. A. (1981), A cortical network for directed attention and unilateral neglect. *Arch. Neurol.,* 10:304–325.

Meyersburg, H. A. & Post, R. M. (1979), An holistic developmental view of neural and psychological processes: A neurobiological integration. *Br. J. Psychiatry,* 135:139–155.

Miles, R. A. (1980), Information processing at the cellular and systems levels in complex organisms. In: *Information Processing in the Nervous System,* ed. A. M. Pinsker & W. D. Willis. New York: Raven Press, pp. 319–329.

Minkowski, M. (1963), On aphasia in polyglots. In: *Problems of Dynamic Neurology,* ed. L. Halpern. Jerusalem: Department of Nervous Diseases, Hadassah University Hospital and Hebrew University Medical School, pp. 119–161.

Mishkin, M., Malamut, R. & Bachevalier, J. (1984), Memories and habits: Two neural systems. In: *Neurobiology of Learning and Memory,* ed. G. Lynch, J. L. McGough & M. Weinberger. New York: Guildford Press, pp. 65–77.

______ & Appenzeller, T. (1987), The anatomy of memory. *Sci. Amer. pp. 80–89.*

Moffett, M. W. (1990), Dance of the electronic bee. *National Geographic, 177,1:134–140.*

Moore, A. Y. (1979), Morphological plasticity of the adult central nervous system.

Scientific Basis of Neurology, Annual Course of American Academy of Neurology, 3212,3:11–18.

Moore, B. E. (1985), On affects: Some biological and developmental perspectives. Unpublished Manuscript.

______ & Fine, B. D. (1968), *A Glossary of Psychoanalytic Terms and Concepts.* New York: American Psychoanalytic Association.

Moreau, T. & Milner, P. (1981), Lateral differences in the detection of touched body parts in young children. *Dev. Psychol.,* 17:351–356.

Morihisa. J. M. & Weinberger, D. R. (1986), Is schizophrenia a frontal lobe disease? An organizing theory of relevant anatomy and physiology. In: *Can Schizophrenia be Localized Within the Brain?* ed. N. C. Andreasen. Washington, DC: American Psychiatric Press, pp. 17–36.

Moruzzi, G. & Magoun, H. J. W. (1949), Brain stem reticular formation and actuation of the EEG. *Electroencephalogr. Clin. Neurophysiol.,* 1:455–473.

Moscovich, M. (1979), Information processing and the cerebral hemispheres. In: *Handbook of Behavioral Neurobiology, Vol. 2: Neuropsychology,* ed. M. S. Gazzaniga. Cambridge, MA: MIT Press, pp. 379–446.

Muller, J. P. (1989), Imre Hermann and modern structural linguistics. Unpublished Paper presented in Budapest, Hungary, November 10, 1989, to the Hungarian Psychoanalytic Society Centenary Conference Honoring Imre Hermann.

______ & Richardson, W. J. (1982), *Lacan and Language.* New York: International Universities Press.

Nadel L. & Wexler, K. (1984), Neurobiology, representations, and memory. In: *Neurobiology of Learning and Memory,* ed. G. Lynch, J. L. McGaugh & N. M. Weinberger. New York: Guildford Press, pp, 125–136.

Nieuwenhuys, R. (1985), *Chemoarchitecture of the Brain.* Berlin: Springer.

Niwa, S. (1989), Schizophrenic symptoms, pathogenic cognitive and behavioral features: Discussion of the 'language of brain' and of 'mind.' In: *Main Currents in Schizophrenia Research,* ed. M. Namba & H. Kaiya. Tokyo: Hesco International, pp. 83–91 [Japanese].

Nottebohm, F. (1985), Neuronal replacement in adulthood. *Annual of New York Academy of Sciences,* 457:143–162

Nunberg, H. (1930), The synthetic function of the ego. In: *Practice and Theory in Psychoanalysis.* New York: International Universities Press, 1960.

Obler, L. K. & Fein, D., eds. (1988), *The Exceptional Brain: Neuropsychiatry of Talent and Special Abilities.* New York: Guilford Press.

Olderfeld-Nowak, B., Ulas, J., Jezierska, M., Skup, M., Wojcik, M. & Domaska-Janik, K. (1985), Role of GML ganglioside in repair processes after hippocampal deafferentation in rats. In: *Central Nervous System Plasticity and Repair,* ed. A. Bignami, F. A. Bolm, C. L, Bolis & A. Adeloye. New York: Raven Press, pp. 85–95.

Ojemann, G. A. (1988), Effect of cortical and subcortical stimulation on human language and memory. In: *Language, Communication, and the Brain,* ed. F. Plum. New York: Raven Press, pp. 101–115.

Olds, J. (1956), Pleasure centers of the brain. *Sci. Amer.,* 195:105–111.

______ (1958), Self stimulation of the brain: Its use in studying local effects of hunger, sex, and drugs. *Science,* 127:315–324.

_______ (1969), The central nervous system and the reinforcement of behavior, *Amer. Psychologist.*, 24:114–132.

_______ & Milner, P. (1964), Positive reinforcement produced by electrical stimulation of septal area and other regions of rat brain. *J. Compar. Physiol. Psychol.*, 47:419–427.

Othmer, E., Hayden, M. P. & Segelbaum, R. (1969), Encephalic cycles during sleep and wakefulness in humans: A 24-hour pattern. *Science*, 164:447–449.

Overstreet, D. H. (1984), Behavioral plasticity and the cholinergic system. *Prog. Neuropsychopharmacol. Bio. Psychiat.*, 8:133–151.

Palay, S. L. & Chan-Palay, V. (1974), *Cerebellar Cortex*. Berlin: Springer.

_______ (1982), *The Cerebellum*. Berlin: Springer-Verlag.

Papez, J. W. (1937), A proposed mechanism of emotion. *Arch. Neurol. Psychiat.*, 38:725–743.

Pearl, J. (1984), *Heuristics: Intelligent Search Strategies for Computer Problem Solving*. Reading MA: Addison-Wesley.

Penfield, W. (1958), The role of the temporal cortex in recall of post experience and interpretation of the present, In: *CIBA Symposium on the Neurological Basis of Behavior*, ed. G. E. W. Wolstenholme & C. M. O'Connor. Boston: Little, Brown, pp. 149–174.

_______ & Mathieson, A. (1974), Memory: Autopsy findings and comments on the role of hippocampus in experimental recall. *Arch. Neurol.*, 31:45–154

_______ & Roberts, L. (1959), *Speech and Brain Mechanisms*. Princeton, NJ: Princeton University Press.

Peon, H., Burst-Carmona, H, Penaloza-Rojas, J. & Bach-y-Rita, G. et al. (1961), The efferent control of the afferent signals entering the central nervous system. *Ann. NY Acad. Sci.*, 89:866–882.

Pfeffer, A. Z. (1963), The meaning of the analysis after analysis. *J. Amer. Psychoanal. Assn.*, 11:229–244.

Phelps, M., Mazziotta, J., Baxter, L. & Gerner, R. (1985), Study design in the investigation of affective disorders: Problems and strategies. In: *The Metabolism of the Human Brain Studied with Position Emission Tomography*, ed. M. Phelps & S. Mazziota. New York: Raven Press, pp. 457–476.

Piaget, J. (1971), *Biology and Knowledge*. Chicago: University of Chicago Press.

_______ & Inhelder, B. (1969), *The Psychology of the Child*. New York: Basic Books.

Piatelli-Palmarini, M., ed. (1981), *Language and Learning: The Debate Between Jean Piaget and Noam Chomsky*. Cambridge, MA: Harvard University Press.

Pinneo, L. R. (1966), On noise in the nervous system. *Psychol. Rev.*, 73:247–270.

Polmin, R. (1990), The role of inheritance in behavior, *Science*, 248–183–188.

Popper, K. R. & Eccles, J. (1977), *Self and Its Brain*. Berlin: Springer.

Premack, D. (1981), Interspecies comparison of cognitive abilities. In: *Language and Learning: The Debate Between Jean Piaget and Noam Chomsky*, ed. M. Piatelli-Palmarini. Cambridge, MA: Harvard University Press.

Pribram, K. (1960), A review of theory in physiological psychology. *Annu. Rev. Psychol.*, 11:1–40.

_______ (1971), *Languages and the Brain*. Englewood Cliffs, NJ: Prentice-Hall.

Pykett, I. L. (1982), NMR imaging in medicine. *Sci. Am.*, 246:78–101.

Racine, R. & Zaide, J. (1976), A further investigation into the mechanisms of the kindling phenomenon. In: *Limbic Mechanisms,* ed. K. E. Livingston & O. Hurnykiewicz. New York: Plenum Press, pp. 457–494.

Rakik, P. (1985), Limits of neurogenesis in primates. *Science,* 2278:1054–1056.

Rayport, S. (1981), Development of the Functional and Plastic Capabilities of Neurons Mediating a Defensive Behavior in Aplysia. Unpublished doctoral dissertation, Columbia University.

Rechtschaffen, A. (1983), General discussion: Dream Psychophysiology. In: *Sleep Disorders,* ed. M. Chase & E. D. Weitzman. New York: Spectrum, pp. 401–413.

Reichert, L. F. (1984), Immunological approaches to the nervous system. *Science, 225:1294–1299.*

Reider, N. (1972), Metaphor as interpretation. *Internat. J. Psycho-Anal.,* 53:463–468.

Reinis, S. & Goldman, J. M. (1982), *The Chemistry of Behavior.* New York: Plenum Press.

Reiser, M. F. (1984), *Mind, Brain, Body: Toward a Convergence of Psychoanalysis and Neurobiology.* New York: Basic Books.

______ (1985), Converging sectors of psychoanalysis and neurobiology: Mutual challenges and opportunity. *A. Amer. Psychoanal. Assn.,* 33:11–34.

Reynolds, G. P. (1987), Dopamine receptor asymmetry in schizophrenia. *Lancet,* April 25, p. 979.

Ricklan, M., Haglin, L., Shulman, M., Cullinan, T, & Cooper, I. S. (1978), Behavioral alterations following acute, short-term, and longer-term cerebellar stimulation in humans. In:*Cerebellar Stimulation in Man,* ed. I. S. Cooper. New York: Raven Press, pp. 161–184.

Rogers, R. (1973), On the metapsychology of poetic language: Model ambiguity. *Internat. J. Psychiat.,* 54:61–74.

Roland, P. E. & Friberg, L. (1985), Localization of cortical areas activated by thinking. *J. Neurophysiol.,* 53:1219–1243.

Rolls, E. T. (1987), Information, representation, processing and storage in the brain. In: *The Neural and Molecular Bases of Learning. [Dahlem Konferenzen,* 1987], ed. J-P. Changeux & M. Konishi. Chichester: Wiley, pp.503–540.

Rose, S. R. & Orlowski, J. (1983), Review of research on endorphins and learning. *J. Dev. Behav. Pediatr.,* 4:131–135.

Rosen, V. (1977), *Style, Character, and Language,* ed. S. Atkin & M. D. Jucovy. New York: Aronson.

Rosenfeld, I. (1984), Review of *Vision* by D. Marr. *New York Review of Books,* October 11, pp. 53–56.

Rosenzweig, M. R., & Bennett, E. L., eds. (1976), *Neural Mechanisms of Learning and Memory.* Cambridge, MA: MIT Press.

______ (1984), Basic processes and modulatory influences in the stages of memory formation. In: *Neurobiology of Learning and Memory,* ed. G., Lynch, J. L. McGaugh & N. M. Weinberger. New York: Guilford Press, pp. 263–288.

Rothenberg, J. & Lenowitz, H. (1989), *Exiled in the World.* Port Townsend, WA: Canyon Press.

Rothstein, A., ed. (1986), *Reconstruction of Trauma.* New York: International Universities Press.

Routtenberg, A. (1976), Doubts about the role of the locus coeruleus in learning and the phosphorylation mechanism engaged in the cerebellum. *Nature,* 260:78–80.

Rozin, P. (1976), The psychological approach to human memory. In: *Neural Mechanisms of Learning and Memory.* ed. M. R. Rosenzweig & E. L. Bennett. Cambridge, MA: MIT Press, pp. 3–48.

Rubinfine, D. L. (1961), Perception, reality testing, and symbolism. *The Psychoanalytic Study of the Child,* 16:73–89. New York: International Universities Press.

Rubinstein, B. B. (1972), On metaphor and related phenomena. In: *Psychoanalysis and Contemporary Science, Vol. 1,* ed. R. R. Holt & E. Peterfreund. New York: Macmillan, pp. 70–108.

_______ ed. (1973), *Psychoanalysis and Contemporary Science, Vol. 2.* New York: Macmillan.

Ruesch, J. & Kees, W. (1964), *Nonverbal Communication.* Berkeley: University of California Press.

Sackheim, H. A., Gur, R. C. & Saucy, M. (1978), Emotions are expressed more intensely on the left side of the face. *Science,* 202:434–436.

Sacks, O. (1985), *The Man Who Mistook His Wife for a Hat.* New York: Summit.

_______ (1989), *Seeing Voices.* Berkeley: University of California Press.

Saitoh, O., Niwa, S., Hiramatsu, K., Kameyama, T., Rymar, K. & Itoh K. (1984a), Abnormalities in late positive components of even-related potentials may reflect a genetic predisposition to schizophrenia. *Biol. Psychiat.,* 19:293–303.

Saitoh, O., Niwa, S., Hiramatsu, K., Kameyama, T., Rymar, K. & Itoh, K. (1984b), P300 in siblings of schizophrenic probands. In: *Advances in Biological Psychiatry,* ed. J. Mendlewicz, & H. M. van Praag. Basel: S. Karger, pp. 46–59.

Sakai, K. (1984), Central mechanisms of paradoxical sleep. In: *Sleep mechanisms: Experimental Brain Research,* Supp. 8, ed. A. Borbery & J-L. Valatx. Berlin: Springer, pp. 3–18.

Saussure, F. de (1959), The linguistic sign: The object of linguistics. In: *Semiotics,* ed. R. E. Innis. Bloomington: Indiana University Press, 1985, pp. 203–230.

Scheller, R. S., Kaldany, R. R., Kreiner, T., Mahon, A. C., Nambu, J. R., Schaefer, M. & Taussig, R. (1984), Neuropeptides: Mediators of behavior in Aplysia. *Science,* 225:1300–1308.

Schildkraut, J. J. & Kety, S. S. (1967), Biogenic amines and emotion. *Science,* 156:21–30.

Schlessinger, N. & Robbins, F. P. (1975), The psychoanalytic process: Recurrent patterns of conflict and changes in ego functions. *J. Amer. Psychoanal. Assn.,* 23:761–782.

Schlessinger, N. & Robbins, F. P. (1983), *A Developmental View of the Psychoanalytic Process.* New York: International Universities Press.

Schwartz, A. (1983/1985), Not art but science: Application of neurobiology and ethology to psychotherapeutic technique and the understanding of transference. Unpublished manuscript.

_______ (1987), Drives, affects, behavior, and learning: Approaches to a psychobiology of emotion and to an integration of psychoanalytic and neurobiologic thought. *J. Amer.Psychoanal. Assn.,* 35:467–506.

Schwartz, S. (1980), *The Mapping of America.* New York: Abrams.

Sebeok, T. A. (1977), Zoosemiotic components of human communication, In:

Semiotics: An Introductory Anthology, ed. R. E. Innis. Bloomington: Indiana University Press, 1985, pp. 292–324.

Shallice, T. (1988), *From Neuropsychology to Mental Structure.* Cambridge: Cambridge University Press.

Shannon, C. & Weaver, W. (1949), *The Mathematical Theory of Communication.* Urbana: University of Illinois Press.

Shapiro, T. (1971), The symbolic process: A colloquium. *Amer. Imago, 28:195–215.*

Sharpe, E. F. (1940), Psycho-physical problems revealed in language: An examination of metaphor. *Internat. J. Psycho-Anal.* 21:201–213.

——— (1950), *Collected Papers on Psychoanalysis,* ed. M. Brierly. London: Hogarth Press.

Shashoua, V. E. (1985), Biochemical changes in the CNS during learning. In: *The Neural Basis of Behavior,* ed. A. L. Beckman. Jamaica, NY: Spectrum, pp. 139–164.

Shepherd, G. M. (1979), *The Synaptic Organization of the Brain.* New York: Oxford University Press.

Sidtis, J. J., Volpe, B. T., Wilson, D. H. Rayport, M. & Gazzaniga, M. S. (1981), Variability in right hemisphere language function after callosal section: Evidence for a continuum of generative capacity. *J. Neurosci.,* 1:323–331.

Simon, H. A. (1979), *Models of Thought.* New Haven: Yale University Press.

Sklansky, M. (1984), Some observations on learning inhibition in college students: A developmental failure in the course of autonomy. In: *Late Adolescence: Psychoanalytic Studies,* ed. D. D. Brockman. New York: International Universities Press, pp. 213–225.

Slap, J. W. & Trunnell, E. E. (1987), Reflections on the self state dream. *Psychoanal. Quart.,* 56:251–262.

Smith, R. C. (1985), Sleep states and learning: A review of the animal literature. *Neurosci. Biobehav. Rev.,* 9:157–168.

Snyder, S. H, (1984), Neurosciences: An integrative discipline. *Science,* 225:1255–1257.

Speigel, J. & Machotka, P. (1974), *Messages of the Body.* New York: Free Press.

Sperry, R. W. (1968), Hemisphere deconnection and unity in conscious awareness. *Am. Psychol.,* 23:723–733.

——— (1970), Perception in the absence of the neocortical commissures. *Perception & Its Disorders,* 78:123–138.

Spreen, O. (1976), Neuropsychology of learning disorders: Postconference review. In: *The Neuropsychology of Learning Disorders,* ed. R. M. Knights & D. J. Bakker. Baltimore, MD: University Park Press, pp. 445–468.

Springer, S. P. & Deutsch, G. (1981), *Left Brain, Right Brain.* San Francisco: W. H. Freeman.

Spruiell, V. (1990), Book review of "A Phylogenetic fantasy: Overview of the transference neuroses" by S. Freud. *Psychoanal. Quart.,* 59, 1:115–122.

Squire, L. R. (1986), Mechanisms of memory. *Science, 232:1612–1619.*

——— Cohen, N. J. & Nadel, L. (1982), The medial temporal region and memory consolidation: A new hypothesis. In: *Memory Consolidation.,* ed. H. Weingartner & E. Parker. Hillsdale, NJ: Lawrence Erlbaum Associates, pp. 137–142.

Stallcup, W. P., Beasley, L. & Levine, J. (1983), Cell-surface molecules that characterize different stages in the development of cerebellar interneurons. *Cold Spring Harbor Symp. Quant. Biol.,* 48:761–774.

Steiner, R. (1987), Some thoughts on "La Vive Voix" by Ivan Fónagy. *Internat. Rev. Psycho-Anal.,* 14:265–272.

Stern, D. N. (1985), *The Interpersonal World of the Infant.* New York: Basic Books.

Stern. M. M. (1988), *Repetition and Trauma,* ed. L. B. Stern. Hillsdale, NJ: The Analytic Press.

Stern, W. C. (1970), The relationship between REM sleep and learning: Animal studies. In: *Sleep and Dreaming.* ed. E. Hartmann. Boston: Little, Brown, pp. 249–257.

Strachey, J. (1934), The nature of the therapeutic action of psychoanalysis. *Internat. J. Psycho-Anal.,* 1969, 50:275–292.

Sutcliffe, J. G., Milner, R. J., Gottesfeld, J. M. & Reynolds, W. (1984), Control of neuronal gene expression. *Science,* 225:1305–1315.

Szentagothai, J. (1975), The module concept in cerebral architecture. *Brain Res.,* 95:475–486.

Thach, W. T., Perry, J. G. & Schieber, M. H. (1982), Cerebellar output: Body maps and muscle spindles. In: *The Cerebellum,* ed. S. L. Palay & V. Chan-Palay. Berlin: Springer, pp. 440–453.

Thompson, R. F. (1975/1977), *Introducción a la Psicológia Fisiológica.* Cap. VI. México: Harla.

_______ (1986), The neurobiology of learning and memory. *Science.* 233:941–947.

_______ (1987), Activity-dependence of network properties. In: *The Neural and Molecular Basis of Learning. [Dahlem Konferenzen],* ed. J-P. Changeux & M. Konishi. Chichester. NY: Wiley, pp. 473–502.

Tobler, I. (1984), Evolution of the sleep process: A phylogenetic approach. In: *Sleep Mechanisms: Experimental Brain Research,* Suppl. 8., ed. A. A. Borbely & J-L. Valatx. Berlin: Springer, pp. 207–226.

Tolpin, M. (1983), Toward the metapsychology of injured self-cohesion. Presented to meeting of the Chicago Psychoanalytic Society, Feb. 22.

_______ & Kohut, H. (1980), The disorders of the self: The psychopathology of the first years of life. In: *The Courses of Life,* ed. S. I. Greenspan & G. H. Pollock. Washington, DC: NIMH, pp. 425–442.

Tolpin, P. (1983), Self psychology and the interpretation of dreams. In: *The Future of Psychoanalysis,* ed. A. Goldberg. New York: International Universities Press, pp. 255–271.

Tompkins, S. (1962a), *Affect, Imagery, and Consciousness, Vol I.* New York: Springer.

_______ (1962b), *Affect, Imagery, and Consciousness, Vol II.* New York: Springer.

Trevarthen, C. (1979), The tasks of consciousness: How could the brain do them? *Mind and Brain.* CIBA Foundation Series 69,, Excerpta Medica, pp. 187–253.

_______ (1985), Facial expressions of emotion in mother–infant interaction. *Hum. Neurobiol.,* 4:21–32.

Tsunoda, T. (1987), *The Japanese Brain.* Tokyo: Daishu Shoten [Japanese].

Tucker, D. M., Watson, R. T. & Heilman, K. M. (1977), Discrimination and evocation of affectively intoned speech in patients with right parietal disease. *Neurology,* 27:947–950.

Tulving, E. & Schacter, D. L. (1990), Priming and human memory systems. *Science,* 247:301–306.

Utena, H. (1979), Episodic memory and the group of functional disconnection states: The biological understanding of schizophrenia. *Psychiatr. Med.,* 21:543 [Japanese].

______ (1984), The reconstruction-of-life therapy method. *Psychiatr. Med.,* 26:803 [Japanese].

Valenstein, A. F. (1961), The psychoanalytic situation. *Internat. J. Psycho-Anal.,* 43:315–324.

Valenstein, E. S. (1968), Modification of motivated behavior elicited by electrical stimulation of hypothalamus. *Science,* 15:1119–1121.

Varon, S. & Manthrope, M. (1985), In vitro models for neuroplasticity and repair. In: *Central Nervous System Plasticity and Repair,* ed. A. Bignami, F. A. Bolm, C. L. Bolis & A. Adeloye. New York: Raven Press, pp. 13–23.

Vartanian, G. A. (1985), Memory function and recovery after brain lesion. In: *Central Nervous System Plasticity and Repair,* ed. A. Bignami, F. A. Bolm, C. L. Bolis & A. Adeloye. New York: Raven Press, pp. 24–34.

Von Bertalanffy, L. (1968), *General Systems Theory.* New York: Braziller.

Von Neumann, J. (1967), *The Computer and the Brain.* New Haven, CT: Yale University Press.

Vygotsky, L. S. (1934), *Thought and Language,* ed. & trans. E. Hermann & G. Vakar. Cambridge, MA: MIT Press, 1962.

Wada, J. A. & Davis, A. E. (1977), Fundamental nature of human infant's brain asymmetry. *Can. J. Neurosci.,* 4:203–208.

Walker, S. F. (1980), Lateralization of function in the vertebrate brain: A review. *Br. J. Psychol.,* 71:329–367.

Washburn, S. L. & Avis, V. (1958), Evolution of human behavior. In: *Behavior and Evolution,* ed. A. Roe & G. G. Simpson. New Haven: Yale University Press.

Washburn, S. L. & Harding, R. S. O. (1975), Evolution and human nature. In: *American Handbook of Psychiatry,* Vol 6: *New Psychiatric Frontiers.* ed. D. A. Hamburg & H. K. H. Brodie. New York: Basic Books, pp. 3–13.

Washburn, S. L., Hamburg, D. A. & Bishop, N. H. (1974), Social adaptation in nonhuman primates. In: *Coping and Adaptation,* ed. G. V. Coehlo, D. A. Hamburg, & J. E. Adams. New York: Basic Books, pp. 3–12.

Wasserman, M. D. (1984), Psychoanalytic dream theory and recent neurobiological findings about REM sleep. *J. Amer. Psychoanal. Assn.,* 32:831–846.

Weinberger, D. R., Berman, K. F. & Zec, R. F. (1986), Physiological dysfunction of dorsolateral prefrontal cortex in schizophrenia I: Regional cerebral blood flow (rCBF) evidence. *Arch. Gen. Psychiat.,* 43:114–124.

Wells, G. A. (1987), *The Origin of Language.* LaSalle, IL: Open Court.

Wernicke, C. (1874), *Der Aphasische Symptomenkomplex.* Breslau: Cohn & Weigert.

Wexler, B. E. (1980), Cerebral laterality and psychiatry: A review of the literature. *Amer. J. Psychiat.,* 137:279–291.

Winson, J. (1985), *Brain and Psyche.* Garden City, NY: Anchor Press.

Wolff, P. H. (1967), Cognitive considerations for a psychoanalytic theory of language acquisition. *Psychological Issues,* Monogr. 18/19. pp. 300–343. New York: International Universities Press.

Yakovlev, P. I. & Lecours, A. R. (1967), The myelogenetic cycles of regional maturation of the brain. In: *Regional Development of the Brain in Early Life,* ed. A. Minkowski. Oxford: Blackwell, pp. 30–70.

Young J. Z. (1978), *Programs of the Brain.* New York: Oxford University Press.

Zaidel, D. & Sperry, R. W. (1974), Memory impairment after commisurotomy in man. *Brain,* 97:263–272.

Zetzel, E. (1956), Current concepts of transference. *Internat. J. Psycho-Anal.,* 37:369–376.

Index

A

Abreaction, 6
Abstract code, bridging hemispheres, 71
  See also Hemispheres; Cerebellum
Abstraction formation, 157, 159, 193, 213
"Action language," *see* Schafer, R.
Adams, R. B., 187, *225*
Adaptation (adaptive decision making), 98, 103, 109
  See also Neural control
Adeloye, A., *227*
Adler, A., 106
Affectivity, 1, 46, 8–13, 122, 194, 212
  and cerebellar control, 65
  metaphorical language, and, 8
  of analyst, 8–13
  See also Limbic system
Agnati, L. F., 48, *225*
Agranoff, B. W., 199, *225*
Aguayo, A. J., 46, *225*
Akimoto, M., 91, *225*
Albus, J. S., 47, 61–62, *225*
Alexander, T., 62, *225*
Alkon, D. L., 47, 50, *225*
Alliance, *See* Therapeutic alliance
Almansi, R. J., 55, *225*
Alzheimer's disease, 188
  See also Basal Nucleus of Meynert
Altman, L. L., 130, 171, *225*
Alverson, H., 205
AMESLAN,, 201
  See also Deafness
Amnesic syndromes, 59
  See also Memory
Amodal perception, 159
  See also Intersensory integration
Amygdala, 188, 193, 195
Analyzability, and brain plasticity, 21
Anderson, J. A., 62, *225*
Andreasen, N., 86, 95, 112, *225*
Annett, M., *226*
Anthropoids, 207
Antrobus, J. S., 139–140, *230*
Anzieu, D., 101, *226*
Aphasia, 18–19, 24–28, 34–39
Aplysia, 49
  See also Kandel, E.
Appenzeller, T., 188
Archaic imago, 2

Arbib, M. A., 60, *226*
Arieti, S., 3, *226*
Aristotle, 22, 28, 202
Arkin, A. M., 124, *226*
Arlow, J. A., 3, *226*
Armitage, R., 67–68, 236
Arrests in development, 2, 124
Artificial intelligence, 61–62
Aschoff, J., 62, 68, *226*
Aseriality, *See* Speech, aseriality
Aserinsky, E., 131
Asymmetry (hemispheric)
  adaptive advantage of, 40
  brainstem, 69
  Dax's contribution, 18, 19
  general review, 17–22
  See also Broca; Hemispheres
ATP, 50
Attachment, therapsid vs. reptile, 205–207
Attention (selective), 1, 12–13, 15, 86–87, 101. Ch. 4 (esp. 89–93)
Attunement, 212
Australopithecus, 20, 39, 207–208, 210
  A. aferensis, 207
  A. boisei, 207
Autonomy, primary vs. secondary, 64
Auxilliary superego, 2
Avis, V., *246*

B

Bachevalier, J., *239*
Bach-y-Rita, P., 46, 101, 226, *241*
"Backup," 74–76, 192
Bakan, P., 135, 138, 142, *226*
Bard, P., 226
Barr, D. L., 95, *239*
Bartus, R. T., 58, *226*
Basal ganglia, 96, 191, 222
Basal Nucleus of Neynert, 188
Basic rest-activity cycle (BRAC), 68
Basch, M. F., 9, 11–12, 32, 41, 108, 126, 134, 147, 149, 187–188, 190, *226*
Basho, Matsuo, 162
Bateson, G., 203–204, *227*
Baudry, M., 50, *238*
Baxter, L., *241*
BEAM studies, 95
Beasley, L., 47, *244*
Beer, B., *226*
Bellugi, U., *227*
Benedek, T., *227*
Bennett, E. L., 46, 150–151, 156, *242*
Benson, D. F., *227*
Benvenuto, B., 162, *227*
Berger, L., 130–132, *227*
Berger, R. J., *227*
Bergmann, M. S., 176, *229*
Berman, K. F., 95, *227, 246*
Bernheim, H., 14, *230*
Bertalanffy, L., *see* von Bertalanffy, L.
Bertini, M., 138, *227*
Bharucha, J., 115
Bibring, E., 6–7
Bignami, A., 47, *227*
Binswanger, L., 18, 26
Bishop, N. H., 217, *246*
Black, I. B., 44, 55, *227*
Black, M. M., 45, *227*
Blair, N., *235*
Bloom, F. E., 44, 56, *227*
Blum, H., 55, *227*
Boden, M. A., 110, 159, *227*
Bolis, C. L., *227*
Bolm, F. A., *227*
Borberly, A. A., 141, *227*
Boschan, P. J., 101, *227*
Bower, G. H., 110, *227*
Brainstem, 56–57, 59, 69, 90, 96, *191*
Breuckner, D. A., 173, 185
Breuer, J., 13, *227*
Bridging
  "horizontal and vertical," 1, 4, 9–11, 14
  of time, 1, 4, 11, 71
  See also Metaphor
Broca, P., (Broca's area) 19, 49–50, 98, 114, 161, 197, 208
Brockes, J. P., 46, *228*
Broughton, R. J., 122, *228*
Brown, J., 19, 25, 26, *228*
Brownstein, M. J., 50, *236*

Brozoski, T., 89, *228*
Brust-Carmona, H., 101, *241*
Buchsbaum, M. S., 95, *228*
Bugbee, N. M., *233*
Bühler, K., 202–203, *228*
Burghardt, G. M., 206–207, *228*
Byrne, J., 122, 124, *230*

C

Calcium channels, 45, 50
Call, J., *228*
Callaway, E., 101, *228*
Calmodulin, memory mechanism, 50
  See also Memory
Cannon, W. B., *228*
Carrol, R. L., 205–207, *228*
Cartesian dualism, 22, 188
Cascade, chemical, 45, 50, 115
Case presentations
  (1) a shooting gesture, 152
  (2) intonation of voice, 152–153
  (3) enactment with total body positioning, 153–154
  (4) a whistling interpretation, 154
  (5) silence can be loud or soft, 154–155
  (6) mixed messages, 155–156
  Mr. D., with dream of composer and lyricist, 2
  Mr. A., father, the tailor, mends, 4–5
  Miss B., wishes to turn off my light, 5–6
  Sperry's patient, 31–32
  Analysand with gun gesture, 32
  Railway conductor with love for French, 35–38
  Mr. B. losing I.Q. points, 50–52
  Mr. H. with learning disability, 52–54
  Mrs. F. and her limb distortions, 54–55
  Maury's dream, 83
  Mr. R.'s disorganization secondary to treatment for brain tumor, 112
  Dr. Z., management of transference with nonverbal aspects, 174–184
Catalepsy, 121
Central parietal cortex, 59, 63, 192
Cerebellar-corticonuclear microcomplex, 46
  See also Cerebellum
Cerebellum, 24, 43, 55–57, 60, 65–72, 188, 191, 196, 222
  anatomy and physiology, 72–74
  blocking of input, 73–74, 194
  body maps, 75, 190
  bridging of cerebral hemispheres, 67–71
  compared with retina, 72
  digital structure, 73
  integration and coordination, 65–72, 74, 191, 195
  neural control, and, 95–97
  nuclei, 60
  storing and restoring programmatic memories, 74–77
  See also Cerebellar-corticonuclear microcomplex; Self; Learning; Memory; Vestibulo-cerebellar system
Chan-Palay, V., 46, 72, 75–76, *241*
Charniak, E., 62, *228*
Charcot, J. M., 34
Chino, Y. M., *235*
Chomsky, N., 97, 155, 161, 165, 202–204, 209, 214–217, *228*
Chugani, H. T., 191, *228*
Chuong, C., 47, *230*
Circadian oscillator, *see* Basic rest-activity cycle, Cycles; Entrainment, Suprachismic nucleus
Clancey, W. J., *228*
Cleghorn, R. A., 57, *228*
Cohen, P. R., 62, *228*
Cohesiveness, 190–192, 194
  See also Self
"Collective unconscious," 133
Collins, 111, 228
Commissures,
  anterior, hippocampal, posterior, 21
  See also corpus callosum
Communication, 145–146, 153, 204, 223

clinical vignettes of modes, 152–156
classification of modes, 150–152
idiot savant use of prime numbers, 150
insects (bees, grasshoppers), 157
mixed messages, 135–156, 161, 192
modes, general, 146–148
nonverbal, 147–149
relationship to personality, 151–152
Shannon's theory of, 6
theoretical discussion, 156–160
See also "Gematria"; "Notarikon"; "Temura"
Condensation, 146, 159–160
Condillac, E. B., 209, 216
Conflict, 18, 22, 41–42, 130
Contiguity vs. similarity, *See* similarity vs. contiguity
Cooper, I. S., 60, *227*
Core self, *See* Self
Corhalles, M., 68, *229*
Corpus callosum, 20, 71
See also Split brain research; Split consciousness
Corticolimbic system, 135–137, 169, 197–198
Corticostriatal system, 135–137, 169
Cotman, C. W., 44–45, 47, *229*
Cowan, W. M., 47, *229*
Creativity, 40
Crossed-sensory integration, *See* Intersensory integration
Crow, T. J., 113, 139, 195, 221, *229*
Cullinan, T., 60, 242
Cycles, 67–68, 130
See also REM/nonREM
See also Basic rest-activity cycle; Circadian cycles
Cyclic-AMP (3',5' cyclic AMP), 45, 50, 115
Cynader, M. S., *239*

D

Darwin, Charles, 147, 229
Davis, A. E., 20, *246*
Dax, M., 18–19, 19, 197
Deafness, and language exposure, 111–112, 158
Dean, R. L., *226*
Decety, J., 96–97, 113, *229*
Declarative learning, 127, 189
See also Learning; Memory
Defense, psychological
possible physiological mechanism(s), 31
See also Disavowal; Repression
DeLisi, L. E., 95, *228*
Dembo, D., 101, *228*
de Medici, Lorenzo, 219
Democritus, 22
Demos, E. V., 187, *229*
DeRobertis, E. M., 109, *229*
Derrida, J., 107, 148, 214
de Saussure, F., *see* Saussure, F.
Descartes, Rene, 22
Desmedt, J. E., 132, *229*
Deutsch, G., 18–19, *244*
Developmental level, and nightmares, 124
DeVoogd, T., 22, 44, 47, *229*
Dewan, E. M., 132–133, 135, 140, *229*
Diencephalon, 58
Disavowal, 31–33, 37–39, 145, 194
Discursive language, 11, 32, 203
See also Symbolism
Displacement, 144, 159–160
Domaska-Janik, K., *240*
Dominance (hemispheric), 76, 138
Donchin, E., 91, *229*
Dooling, E. C., 191, *232*
"Downloading" of information subsystems, 135–138, 142, 222
Drachman, D. A., 57, *229*
Dreams, 130–133
self-state dreams, 134
right hemisphere activation (and REM), 138–139
See also REM/nonREM
Dreyfus, C. F., 58, *227*
Drive-defense theory, 106
Dual processing, 66–67, 81
Dualism (vitalism), *See* Cartesian dualism

Duncan, G. C., 95, *239*
Dushenko, T. W., 141, *229*

E

Eccles, J., 19, 22, 65–66, 72–73, 75–76, *229, 230, 241*
Edelman, G. M., 47, 47, *230*
Edelson, M., 148, 161, 163, *234*
Edwards, A., 122–124, *230*
Ego
  archaic schemes of, 172
  ego psychology, 107
  Freud's pre- and post-1937 views on, 23
  heritary dispositions of, 21
  observing vs. experiencing, 8
  regression of, 124
Ehrlichman, H., 139–140, *230*
Einstein, Albert, 18, 167, 220, *230*
Embde, R., 136–137, *230*
Emotions,
  and the left side of the face, 27
Empathy, 66
Endicott, K., 45, *234*
Endorphins, 57
Eneuresis, 123
Enkephalins, 46
Entrainment, 141
Erikson, E., 107
Ethology, 61–63
  See also Learning
"Eurisko," 62
Evoked response potential(s), 113, 132
  dominant vs. nondominant hemisphere, 20
Evolution, 126
Eyefields, *see* Frontal eyefields
Exhibitionists, and EEG abnormalities, 196

F

Fairbairn, W. R. D., 107
Fawcett, J. W., *229*
Feigenbaum, E. A., 62, *228*
Fein, D., *240*
Feldman, C. F., 9, 148, *230*
Feldman, S. S., *230*
Ferenczi, S., 166, *230*
Fine, B. D., 29, *240*
Fine, H. J., 3, 85, 170, 230
Fishbein, W., 171
Fisher, C., 68, 122–123, 123–125, 230–231, *235*
"Fixation" vs. "kindling," 187
Fleck, P. A., 68, *236*
Flor-Henry, P., 138, 194–195, 198, *230*
Fónagy, E., 147–149, 160, 162, *230*
Forest, D. V., 3, *230*
Foulkes, D., *230*
Foveal pursuit system, 63–64
"Fragmentation," 68, 128, 195
"Framing," 88
Franzen, G., 95, 97, *234*
Fraser, J. G., 147, 160, *230*
French, T. M., 120, *230*
Freud, Anna, 6, 106–107
Freud, Sigmund, 22–23, 25, 33, 36, 40, 66–67, 79, 81, 99, 100, 103, 107, 130–131, 144, 159–162, 165–166, 169, 194, 207, 223–224
  1888, "Preface to translation of Bernheim's 'Suggestion,' " 14, *230*
  1891a, "On the interpretation of aphasias," 19, *230*
  1891b, "On aphasia," 19, 34, 45, 107, 148, *230*
  1893;-1895 (Breuer and Freud), "Studies on hysteria," 18, 227
  1895, "Project for a scientific psychology," 190, *231*
  1897, "Letter to Fliess," 13, *231*
  1900, "Interpretation of dreams," 83, 130, 148–159, *231*
  1901, "The psychopathology of everyday life," 148, *231*
  1905, "Jokes and their relation to the unconscious," *231*
  1911, "Formulations on the two principles of mental functioning," *231*
  1912, "Totem and taboo," 213–214, *231*

1914, "On narcissism," 106, *231*
1915, "The unconscious," 156, *231*
1914, "The Moses of Michaelangelo," 148, *231*
Friberg, L., 133, 133, *231, 242*
Frick, R. B., 23, 34, 42, 65, 67, 67, 72, 137, *231, 234*
Friedman, D., *226*
Friedman, S., *231*
Fromkin, V. A., *231*
Fromm, E., *231*
Frontal cortex, *see* Prefrontal cortex
Frontal eyefields, 89
Fukuda, M., 91, 92, *231*
Funt, D., *231*
Furer, M., 153
Fuster, J. M., 98, *231*
Fuxe, K., 48, *225*

G

Gabel, S., 130–131, 134
Galaburda, A. M., 40, *231*
Galen, 22
Galin, D., 19, 23, 27–28, 30–32, 41–42, 68–69, *232*
Gall, F., 19, 49, 232
Gangliosides, 46
Gans, E., 214, 232
Gardner, A., 140
Gasanov, U. G., 46, *232*
Gastaut, H., 122–123, *232*
Gaze, in prefrontal cortical injury, 88–89
See also Selective attention
Gazzaniga, M., 19, 46, 60, *232, 237, 244*
Gedanken experiment, 167–168
Gedo, J., 7, 13, 28, 57, 98, 103–111, 115, 117, 147, 154, 169, 187, 195, *232*
"Gematria," 150
See also Communication, modes
Genome, 45, 109, 126
Gerner, R., *241*
Geschwind, N., 19, 59, 63, *232*
Gesturing
See Communication, nonverbal; Language
Gibson, W. C., 18–19, *232*
Gilbert, P., 56, *232*
Gill, M. M., 170, *232–233*
Gilles, F. H., 188, *233*
Giora, Z., 172, *233*
Girard, R., 214
Glial growth factor, 46
Goldberg, A., 28, 105–108, 169, 187, 232
Goldman, A. I., 159, *233*
Goldman, J. M., 199, *242*
Goldman-Rakik, P. S., 87, 113, *233*
Golgi cells, 96
Gomperz, H., 202
Goodall, J., 209–210, *233*
Gottesfeld, J. M., *245*
Gould, S. J., 202, *233*
Grammar, shared by "mind" and brain, 94, 94
Granule cells, of cerebellum, 72, 96
Greenberg, R., 130, 131–132, *233*
Greitz, T., 51, *233*
Grinker, R. R., Sr., 107, 135, *233*
Gross, G., 141, *233*
Grotstein, J. S., 31, *233*
Grubich-Simitis, I., 166
Gunther, M., 187, *233*
Gur, R. C., 27, 68, 69, *233, 241*
Gur, R. E., 68, 69, *233*
Gutwein, B. M., 170

H

Habit pattern learning, 137
See also Learning
Habituation, 49
Hadley, J. L., 194, 195, 196, 198, *233*
Hagan, S., 44, *233*
Haglin, L., 60
Hagman, R. S., 205, *233*
Hamburg, D. A., 207–208, 209–210, 215, 217, 231, *233*, 244, *246*
Hamburger, V., 44
Hamilton, N. G., 67, *233*

Hand gesturing, *See* Gesturing; Signing
Harding, R. S. O., 207, 209, *244*
Hartmann, H., 23, 61, 64, 79, 100, 106, 221, *233–234*
Hayashida, S., 91, *231*
Hayden, M. P., 170
Heath, R. G., 60, *234*
Hebb, D. O., 221, *234*
Heidegger, M., 160
Heilman, K. M., 24, *245*
Heinrichs, J., 115, *234*
Held, R., 223
Hemispheres of the brain, 14, 18, 70, 196–197
  abstract code, 71
  activation of, and head turning, 30
  cycling of activity, 33, 140
  dominance, 20, 68
  effect of psychoanalysis on, 19, 29–33
  integration (collaboration) of, 26–28, 70, 78, 138, 194
  left hemisphere, 138
  right hemisphere, 138, 194, 201
  stereotypes, 24–28, 138
  See also Disavowal; Repression; Speech
Henderson, J. L., 131, *234*
Herder, J. G., *See* von Herder, J. G.
Hermann, 147–148, 160, *234*
Hewes, G. W., 211, *234*
H. erectus, 207–208, 210–211, 213
H. habilis, 207–208, 210, 213
H. sapiens (early), 209–210, 213
Hierarchical (developmental) model, 105, 108–111
  and language level, 117
  and neural control, 109–112
  and prefrontal cortex, 112–114
  See also Language; Neural control; Prefrontal cortex
Hierarchical organization (of mind and brain), 9–11, 14, 50, 59, 117, 189
Hill, D., 22, *234*
Hinde, R., 202
Hippocrates, 22
Hillyard, S. A., 19, 133, *234*
Hippocampal commissure, 21
Hippocampus, 56, 58, 60, 132, 188, 192
  See also REM/nonREM
Hiramatsu, K., 91, 91, 94, 221, *231, 234, 243*
Hobbes, Thomas, 22
Hobson, J. A., 133, 141, *234*
Hoffman, S., 47, *234*
Hoit, M., 64, *234*
Holcomb, H. H., 95, *228*
Holloway, R., 208
Holton, G., 167–168, *234*
Homeobox, 109, 116
Hominids, 209
  See also Australopithecus; A. afarensis; A. boisei; H. habilis; H. erectus; Neanderthal man
Hominoids, 209
Hubel, D. H., 215
Hypothalamic-hypophyseal control, 142

I

"Ideational mimetics," 66
"Imagination set," 92–93
"Individuation," 131
Inferior olivary nucleus, 72
Information-processing model, 69–70
  and REM/nonREM periods, 125
  See also REM/nonREM
Ingvar, D. H., 1, 12, 29, 51, 86, 95, 96–97, 98, 112–113, *229, 234, 237*
Inhelder, B., 191, *241*
Innis, R. E., 202, *234*
Interpretation
  language of, 2, 3
  mechanism(s), 29, 48
  metaphorical transference interpretation, 2, 4–6, 12–23, 29, 30–33
  "mutative" interpretation, 2
  preparation for, 15, 57
  receptivity for, 2
  timing, 164

Intersensory integration, 2, 118, 144–149, 158–160, 191, 192, 212, 221–222
Intonation, *See* Sonority
Introspection, 7
Isaac, G., 209
Isay, R. A., 144, *234*
Isomorphism, 19–20
Isseroff, A., *233*
Itoh, K., 91, *231, 234, 243*
Itoh, M., 43, 65, 65, 72, 74–77, 78, 87–88, 95–97, 99, 115, 117, 196, *230, 234, 235*
Iwanami, A., 91, *231*

J

Jackson, J. H., 8, 28, 59, 111, *235*
Jackson, R., *235*
Jakobson, R., 107, 144–148, 159, 161, 202–203, 212, *235*
James, William, *235*
Japanese language, 84, 117–119
Jeannerod, M., 141, *235*
Jezierska, M., *240*
Jonakait, G. M., *227*
Joseph, E., 188
Joseph, R., 48, *235*
Jucovy, M. E., 176, *229*
Judgment, 85, 90–91
Jung, K. G., 106, 130–131, 133

K

Kass, J. H., 191, 196, *235, 238*
Kahn, E., 122, 124, *235*
Kaldany, R. R., 49, *243*
Kameyama, T., 91, *231, 233, 243*
Kandel, E., 45, 47, 49–50, 58, 187, *235*
Katz, D. M., *227*
Kayton, L., 103, *236*
Kees, W., 147, *243*
Kemper, T. L., *231*
Kennedy, R., 162, 164, *227*
Kent, E. W., 61, 71, 84, 86–91, 99, 113–115, 158, 189–190, 192–193, 295–296, *235*
Kernberg, O., 103
Kety, S., *235*
Kilmer, W. L., 60, *226*
"Kindling," and "fixation," 187
Kinsbourne, M., 30, 69, *235*
Kitney, R. I., 57, *236*
Klein, G., 172, *236*
Klein, M., 107–108, *236*
Klein, R., 67–68, *236*
Kleitman, N., 68, *236*
Kluver-Bucy syndrome, 196
Koh, S. D., 103, *236*
Kohut, H., 7, 21, 108, 125, 128, 133–134, 133, *236, 245*
Kolata, G., 44, 47, *236*
Kolb, B., 22, 87, 113, *236*
Korsakoff's syndrome, 59
Kreiger, D. T., 50, *236*
Kreiner, T., 49, *243*
Kripke, D. F., 68, *236*
Kris, E., 23, 100, *233, 236*
Krubitzer, L. H., *235*
Krystal, H., 121

L

Lacan, J., 107, 148, 160, 161–162, 212
Landstrom, A. L., *235*
Lange, C. G., *236*
Langer, S. K., 10, 203, *236*
Language, 157–158, 202–203
  and regional cerebral blood flow studies, 114
  and communication, 30
  and "control set," 92–95, 104
  and deafness, 111–112, 147, 158
  evolution, 205–211
  gesture, *See* Communication, nonverbal
  language exposure, 111–112, 158
  "langue" vs. "parole," 203
  mime, *see* Communication, nonverbal
  natural role of, 114
  personal language, 154

phonemic analyzing ability, 39
recovery in poplyglots, 34–39
serial aspects, 97–98
symbolization and meaning, 40
Lashley, K. S., 19, 40, 58, *236*
Lassen, N., 1, 12–14, 29, 97, 133, *237*
Lateralization of audition, 94, 114–115
Lavie, P., 139, *237*
Leakey, R. E., 206–208, 235, 208–210, *237*
Learning (knowledge), 46–47, 50, 77–79, 135
and cerebellum, 60
artificial intelligence,
blocks, 134, 196
cholinergic system, 57–58
"declarative" learning, 115
disability (LD), 43, 63–65, 79
"discriminative," 49
endorphins, 57
ethology, 63–65
flatworms, 49
"habit patterns," 115, 138
major subsystems, 135–138
"modules," 61
neurophysiology-neuroanatomy, 58–61
"procedural" learning, 115
relation to brain plasticity, 48–49
REM deprivation, 140
transfer by chemical means, 55–58
trauma, 126
See also Cerebellum; Corticolimibic system; Corticostriatal system; Gaze; Memory; Plasticity; Prefrontal cortex; Vestibulo-cerebellar system
Leavy, S. A., 3, 161, *237*
Lecours, A. R., 190, *246*
LeDoux, J. E., 24, *237*
Lehtonen, J., 18, *237*
Leibniz, G. W., 22
Leider, J., 147, *237*
Leisman, G., 63
LeMay, M., *231*
Lenat, D., 62
Lenowitz, H., *242*
Levels of brain organization, 48
See also Plasticity; Learning; Memory
Levey, M., 237
Levi-Montalcini, R., 44
Levin, F. M., 5, 57, 103, 108, 111, 113–114, 117, 139, 187, 190, 195, 197, 237–238
Levine, J., 47, *244*
Leviton, A., 188, 232
Levy, J., 19–20, 39, 40, 109, *238*
Lewin, B., 3, *238*
"Liaison brain," 20, 22
Libido theory, 107
Lichtenberg, J., 106, 187, *238*
Lichtheim, R., 25
Lieberman, P., 118, 157, 215, *238*
Lilleskov, R. K., 149, 151, *238*
Limbic system, 28, 47, 58–59, 68, 78–80, 90, 138, 191, 193, 197
See also Corticolimbic system
Linguistics, 161–164
Litowitz, B. E., 3, 212, 214–215, *238*
Litowitz, N., 212, 214–215, *238*
Locke, S., *238*
Locus coeruleus, 80, 200
Loewenstein, R. M., 23, *233*
Lorenz, K., 62
Lowenthal, F., 211, *238*
"Lucy" (A. aferensis), 207
"Lucy R" (Freud's case), 13
Luria, A. R., 59, 98, 117, 192
Lusek, C. G., 196, *238*
Lynch, G., 50, *238*

M

"Machine language" of brain, 190
Machotka, P., 148, *244*
MacLean, P. D., 28, 197, 206, *238*
See also Triune brain theory
Magoun, H. J. W., 142, *240*
Mahler, M., 147
Makara, G. B., 142
Mahon, A. C., *243*

Mahony, P., 107
Mammalian evolution from reptiles, 205–206
  See also Therapsids
Maps in the brain, 192–193, 196
Martin, J. B., 50, *236*
Masturbation, 87
Mandelbrot, B., 116
Manning, A., 62–63, *239*
Manthrope, M., 45, *246*
Markey, K. M., 225
Marr, D., 59, 61, 73, 193, *239*
Martin, J. B., 50, *236*
Masochism, 103
Massion, J., 74, *239*
Mathew, R. J., 95, *239*
Mathieson, A., 60, *241*
Muary's dream, 83
Mazziotta, J., *241*
McCallum, W. C., 91, *229*
McCarley, R. W., 133, 141, *234*
McClintock, B., 45, 128, *239*
McConnel, J. V., 49, *239*
McDermott, D., 62, *228*
McGlone, J., 19, *239*
McLaughlin, J., *239*
Memory, 48, 87, 189
  affective organization of, 132
  and psychoanalytic models, 2
  calmodulin, 50
  chemical "cascade," 49–50
  declarative, 189
  episodic, 32–33, 87, 126
  evocative recall, 12
long-term vs. short-term, 45, 49, 59, 102–103, 193
  multiple storage sites, 60, 188
  piracetam effect, 57–58
  procedural, 189
  restoration, 60
  retrieval, 59–60
  scopolamine effect, 57–58
  semantic, 33, 87, 126
  sensory-modality organization of, 9
  sensory stage of, 59
  visual, 18
  See also Amnesic syndromes; Amygdala; Cerebellum; Diencephalon; Hippocampus; Korsakoff's syndrome; Learning; Plasticity
Mental "snapshots," 90, 99, 102, 104
Merzenich, M. M., 44, 49, 58, 63, 191, 223, *239*
Metaphor
  and functional bridging, 9–12, 14, 29
  Freud's usage, 4
  in interpretation, 2, 14
  overcoming repression and disavowal, 29
  See also Transference interpretation
Meyersburg, H. A., 19, 187, *239*
Miall, R. C., 57, *236*
Miles, R. A., 63, *239*
Milner, P., 25, *239, 245*
"Mind," 21, 38, 92
Mime, *See* Communication, nonverbal
Minkowski, M., 18, 34–39, *239*
Mishkin, M., 8, 137, 138, 188, *239*
Modular systems of the cortex, 166
Moffett, M. W., 157, *239*
Monoamines, 56, 58
  See also Dopamine; Serotonin
Monod, J. E., 126–128
Moore, A. Y., 44, 85, *239*
Moreau, T., 25, *240*
Morihisa, J. M., 95, *240*
Morrow, D. G., 110, *227*
Moruzzi, G., 142, *240*
Moscovich, M., 69–70, *240*
Mossy fibers of cerebellum, 72
Mullaney, D. J., 68, *236*
Muller, J. P., 147–148, 160, *240*
Myelinization, 48, 165

N

Nadel, L., 116, *240*
Nakogome, K., 91, *231*

Naloxone, 57
Nambu, J. R., *243*
"Naming," 158
  See also Abstraction formation; Communication; Intersensory integration; Language
Narcissistic injury, 107
Natural language, 116
  See also Language
Nauta, W. J. H., 197
Neanderthal man, 207
jNeocortex, 56
Nerve growth factor (NGF), 44–45
Neumann, J., *See* von Neumann, J.
Neural cell adhesion molecules (N-CAMs), 47
Neural control, 65–71, 84–87, 103–104
  and cerebellum, 95–97
  and prefrontal cortex, 92–95
  feedback, 84–85
  feedforward, 84–85
  of serial aspects of language and speech, 97–98, 114
Neural injury, recovery mechanisms, 44–46
Neurite-promoting factors, 45
Neurogenic shock, *see* Shock of central nervous system origin
Neuroglia, functional anatomy, 45–46
Neuropeptides, *see* Neurotransmitters
Neurotransmitters, 46, 55, 57
Nieuwenhuys, R., 199, *240*
Nieto-Sampedro, M., 44–45, 47, *229*
Nightmares/night terror (pavor nocturnus), 122–123, 125, 142
  relationship to psychological developmental, 124–126
Niwa, S., 84, 86, 91, 92, 93, 94, 99, 104, 113–115, 117, 119, 222, 231, 234, *240, 243*
N 100, 91
Noradrenaline/norepinephrine, 96
  See also Monoamines
Nonverbal communication, 220–221
  See also Communication, nonverbal
"Notarikon," 150
  See also Communication, modes
Nottebohm, F., 22, 44, 47, *229, 240*
Nunberg, H., 100, *240*

O

Object relations theory, 107
Obler, L. K., *240*
Oculomotor system, 63–64
Oderfeld-Nowak, B., 46, *240*
Oedipal phase, 165, 176
  and interhemispheric transmission, 21, 41
Ojemann, G. A., 161, *240*
Olds, J., *240–241*
O'Learly, D. D., *229*
Oliver, G., 109, *229*
Operating system (O.S.), 92–93
  and "imagination set," 94
  "grammar," 94
  "natural language," 93–94, 104, 114–115, 117
  "operation set" (system control set), 92–93
Operator gene, 50
  See also Cascade, chemical
Optokinetic system, 54
Orlowski, J., 57, *242*
Ornstein, R., 30, *231*
Ornston, D., 107
Othmer, E. G., 170
Overstreet, D. H., 57–58, *241*

P

Palay, S. L., 46, 72, 75–76, *241*
Palkovits, G. B., 142, *238*
Pao, Ping Nie, 107
Papert, S., 215
Papez, J. W., 65, 197, *241*
Parent loss, 176
Parietal cortex, *see* Central parietal cortex
Parkinsonism, 97

Pavor nocturnus, *see* Nightmares/night terrors
Pearl, J., 61, *241*
Pelagius, 22
Penaloza-Rojas, J., 101, *241*
Penfield, W., 19, 58, *241*
Penis envy, 175–176
Péon, H., 101, *241*
Peptides, *see* Neurotransmitters
Perception, relation to semiotics, 200
Perlo, M., 83
Perry, J. G., 17, 72, *245*
Personality, and communicative preferences, 151–152
PET scanning, 133, 191
Pfeffer, A. Z., 33, *241*
Phelps, M. E., 51, 191, *241*
Piaget, J., 1, 9–10, 23, 76, 107, 149, 172, 191, 214–216, *241*
Piatelli-Palmarini, M., 214–215, *241*
Pierce, C. S., 202
Pinneo, L. R., 76, *241*
Piracetam, 57–58
Pitrés's principle, 18, 34, 41
Placebo response, basis for, 57
Plasticity, of brain, 21, 29, 41, 43–46, 63, 76–77, 127–128, 137–138
  relationship to learning, 46–49
Plato, 22, 149
Pleasure/pain, 56
Poincaré, H., 214
Polley, E. H., *235*
Pollio, H. R., 230
Polmin, R., 202, 241
Polzner, A., *227*
Pontine nucleus, 72
Pontog. niculo-occipital (PGO) saves, 141
Popper, K. R., 20, 22, *241*
Post, R. M., 19, 187, *239*
Prefrontal cortex, 83–106, 112–114, 222
  as serial processor, 97–98
Premack, D., 202, 208, 212, *241*
"Prerepresentational self," 136
Presentational symbolism, 203
  See also language
Pribram, K., *241*
Primary association cortex, 9, 70
Primate studies, 208–209
"Priming," of memory, 52, 75, 220
  See also Cerebellum; Memory, procedural; Striatem
Procedural learning, 127, 189
  See also Learning, Memory, "Priming"
Prosody, 173–174
  See also Communication, modes; Sonority
Prostoglandins, 46
Protein kinase, 45
"Pruning" of neurons, 47
Pseudodepression, in prefrontal injury, 87
Psychological development, 185–200
Psychosis, 195
  See also Schizophrenia
Purkinje cells, 46
Purves, D., 44
Pykett, I. L., 29, *241*

R

Racine, R., 44, *241*
Rakik, P., 47, *242*
Randall, J. N., *239*
Rapaport, D., 108, 172
Rayport, S., 49, *242*
Reaction time (RT), in schizophrenia, 91, 94, 113
Rechtschaffen, A., 133, *242*
Recency, 87
  See also Memory
Recouer, P., 107
Red nucleus, 72
Regenerativity of nerves, 45
Regional cerebral blood flow (rCBF), 12, 97, 114, 132–133
Regional cerebral metabolic rate (rCBMR), 97, 114

Regression, 168
in the service of the ego, 100, 124
Reichert, L. F., 55, *242*
Reid, T., 207
Reider, N., 3, *242*
Reinis, S., 197, *242*
Reiser, M. F., 87, 106, 113, 187, *242*
REM/nonREM phenomena, 56, 67–68, 121, 124–125, 126, 129, 132–133, 135, 137, 139–141, 169–171, 222
REM deprivation (RD), 140
Repetition compulsion, 126
Representation, theory, 202–203
Repression, 150, 194
barrier, 194
theory, 2, 31–33, 37–39
Response inhibition, 87
"Response set," 92–93
See also Neural control
Response task delay, 87
Restoring (programmatic) memories, 75–76
Reticular activating system, 89
Retinal slippage, 64
Reynolds, G. P., 198, *242*
Reynolds, W., 198, *242*
"Rhythm," *See* Communication, modes
Richardson, W. J., 160, *240*
Ricklan, M., 60, *242*
Riddell, P. M., 57, *236*
Ritter, W., 91, *229*
Roberts, L., 19, *241*
Robbins, F., 33, 169, *242*
Rogers, R., *242*
Roland, P. E., 133, *231*
Rolls, E. T., 188, 193, *242*
Rose, S. R., 57, *242*
Rosen, V., 107, 148, 159, 161–163, *242*
Rosenfeld, I., 3, 61, 62, *225*, *242*
Rosenzweig, M. R., 46, *242*
Rosvold, H. E., *228*
Rothenberg, J., 150, *242*
Rothstein, A., *242*
Rousseau, J.-J., 209
Routtenberg, A., 56, *242–243*
Rozin, P., 59, *243*
Rubinfine, D. L., 170, *243*
Rubinstein, B. B., 3, 190, *243*
Ruesch, J., 147, *243*
Russell, Bertrand, 212
Rymar, K., 91, *243*

S

Saccadic eye movements, 63
See also Gaze
Sackheim, H. A., 27, *243*
Sacks. O., 111, 114, 147–149, *243*
Saint Augustine, 22
Saitoh, O., 91–92, 93, 231, 233, *243*
Sakai, K., 142, *243*
Sapir, E., 162
Sasaki, K., 74, 91, *239*
Sasaki, T., *231*
Saucy, M., 27, *243*
Saussure, F. de, 107, 144–146, 159–62, 202–204, 212, *243*
Schacter, D. L., 76
Schaeffer, M., *243*
Schafer, R., 148, 161, 163–164
Scheller, R. S., 49, *243*
Schieber, M. H., 72, 75, *245*
Schildkraut, J. J., *243*
Schizophrenia, 91–95, 102–103
and brain language system, 92
event-related potentials, 91
hypofrontality, 95
mechanisms, 94–95
reaction time (RT), 91
scanning research, 95
Schlessinger, H., 149
Schlessinger, N., 18, 33, 52, 169, *243*
Schools of psychoanalysis, 106–107
Schoppmann, A., *239*
Schrödinger, E., 122
Schulman, M., 60, *242*
Schwann cells, 46
Schwartz, A., 106, 187, *243*
Schwartz, S., 219, 243
Scopolamine, 57–58
Sebeok, T. A., 205, *243*
Secondary process, 10

Segelbaum, R., 172
Segelman, R., 172
Selective attention, *See* Attention
Self (core self), 137, 190–192
  self-definition, self-organization, self-regulation, 108
  See cohesiveness, self-in-the-world model
Self-analysis, 162
Self-in-the-world model, 43, 95–96, 99, 110, 137, 195
Selfobject, *see* transference
Self psychology, 106, 128
Self-reflection, in machines, 62
Self-state dreams, 134
Selye, H., 122
Semantic memory (mode), 33, 189
  See also Communication; Language; Memory
Semiotics (semiology), 163–164, 202–205
Sensitization, 49–50, 137
Sensorimotor experience, 11–12
"Serial action program," 97
Seriality (temporality), 114
Serota, H., 152
Serotonin (5-hydroxytryptamine), 49–50, 96
  See also Cascade, chemical
Shallice, T., 112, 159, *244*
Shannon, C., (communication theory), 6, 150
Shapiro, T., 3, *244*
Sharpe, E. F., *244*
Shashoa, V. E., 199, *244*
Shepherd, G. M., 56, 66, 72–73, *244*
Sherman, K. A., *226*
Shiromani, J., 170
Shock, of central nervous system origin, 122–123, 142
Shortliffe, E. H., 62, *228*
Sialic-acid content, 47
Sidtis, J. J., 24, 39, *244*
Signing (sign language), 5, 148–149
  See also AMESLAN
Silence, possible meanings, 154–155
Similarity vs. contiguity, 58, 147, 159, 160, 202
Simon, H. A., 47, 61, *244*
Simplinson, C. H., *230*
Singer, S. P., *244*
"Sixth sense," 151
Skinhoj, E., 1, 12, 29, *237*
Skinner, B. F., 215
Sklansky, M., 64, *244*
Skup, M., *240*
Slap, J. W., 130, 133, 134, *244*
Sleep, 129–130, 135
  sleepwalking (somnambulism), 123
  See also Dreams; Hypothalamic-hypophyseal control; REM/nonREM; suprachiasmic nucleus
Slips (parapraxis), 162
Smith, R. C., 140, *244*
Snyder, S., 46–47, *244*
Somatosensory map, 58
Sonority, 148, 153, 160, 177, 209
Speech
  and hemispheric specialization, 18–19, 24
  aseriality, 97–98
  diminished spontaneity, 86
  speech apparatus, 24–25, *24, 25, 26, 30, 31, 36*
Sperry, R. W., 19, 28, 31–32, *244*
Spiegel, J., *148, 244*
Spinelli, D. N., 60, *226*
Spinoza, B. De, 22
Split-brain research, 31–32
Split consciousness, 24–25, 30–31
Spreen, O., 63, *244*
Springer, S. P., 18–19, *244*
Sprouting, 45
Spruiell, V., *244*
Spurzheim, C., 19, *232*
Squire, L. R., *244*
Stage IV sleep, 123
Stallcup, W. P., 47, *244*
Stanfield, B. B., *229*
Stanford University, 62
Stein, J. F., 57, *236*

Steiner, R., 146, 242
Sterba, R., 7
Sterman, M. B., 171, *229*
Stern, D. N., 145, 159, 187, 191, 212, *245*
Stern, M. M., 121–130, 134, 142, *245*
Stern, W. C., 140, *245*
Stimulus set/response set, 92–93
Strachey, J., 2–3, 7, 15, 29, *245*
Striatem, 186
Structural linguistics, 212
See also Linguistics
Stryker, M. P., *237*
Stuss, D. T., 112, *227*
Suggestion, 6
Sullivan, H. S., 199
Superego, 7
Suproachiasmic nucleus, 141
Suttcliffe, J. G., 45, *245*
Symbolism
discursive, 11, 203
presentational, 11, 203
Synapse(s)
presynaptic membrane chemical cascade, 44–45
role in information processing, 44, 46
Synesthesia, 159
Syntactical memory (mode), 158
See also Communication; Language, Memory
Synthetic activity, 12, 46
Synthetic function, 100
Szentagothai, J., 39, 65, 75–76, 142, *230, 239, 245*

T

Takahashi, T., 67, *233*
Taussig, R., *243*
"Teleonomic principle," 126–127
Temporality, *see* Seriality
Temporal lobe(s), 58–59, 192–193
See also Amygdala; Hippocampus; Kluver-Bucy syndrome
"Temura," 148
See also Communication, modes
Tension, attribution of meaning to states of, 126
Thach, W. T., 72, 75, *245*
Thalamus, 59, 89, 191
thalamocortical activating system, 160–161
ventrolateral nucleus of, 72, 161
Therapeutic alliance, 7
Therapsids, 205–206
Thompson, R. F., 65, 188, *245*
Timbre, *See* Communication; Sonority
Tobler, I., 135, *245*
Tokyo University, 84, 91–92, 95
Tolpin, M., 125, 133–13, *245*
Tolpin, P., 125, 133–134, *245*
Tompkins, S., 145, *245*
Tools, 209–210
Toulmin, S., 9, *230*
Transference, 165–171
as "probe," 171
and REM/nonREM cycles, 140, 170
expanded definition, 7
Freud's phylogenetic view of, *166*
"metaphorical transference interpretation," 2, 14
relation to hemispheres, 30, 44
requirements for appearance of, 172
selfobject transferences, 7
"Transfert," 14
Trauma (biotrauma), 126, 142
Traumatic dreams, 124
See also Nightmare
Trevarthen, C., 39, 137, 153, 187, *238, 245*
Tripartite model (of Freud), 103
Triune brain theory, 43, 197
See also MacLean, P. D.
Trunnel, E. E., 130, 134, *244*
Tsunoda, T., 84, 94, 99, 104, 114–115, 117–118, 222–223, *245*
Tucker, D. M., 24, *245*
Tulane University, 60
Tulving, E., 76, 246

Turing, A. M., 61–62
Tzischinsky, O., 139, *237*

U

Ulas, J., *240*
Ullman, M., 131
Unconscious, 20
Utena, H., 91, *246*
and mime, 149

V

Valenstein, A. F., 122, *246*
Varon, S., 45, *246*
Vartanian, G. A., 46, *246*
Vasopressin, 46
Vergence system, 64
Vespucci, Amerigo, 219
Vestibulo-cerebellar system, 24, 135–137, 169, 191, 195–196
Vestibulo-ocular reflex (VOR), 64, 77
Vicotr, M., 187, *224*
Vitalism (dualism), 22
Vocalization, evolution of human, 205–207, 208
Volpe, B. T., *244*
von Bertalanffy, L., 19, 107, 135, 190, *246*
von Herder, J. G., 209, 215
von Neumann, J., 9, 185, *246*
Vuckovich, D. M., 17–84, 190
Vygotsky, E. S., 211, *246*

W

Wada, J. A., 20, *246*
Waldseemüller, Martin, 218
Walker, S. F., 19, *246*
Washburn, S. L., 207, 209, 217, *246*
Wasserman, M. D., 130, *246*
Watson, R. T., *244*
Weaver, W., *243*
Weinberger, D. R., 95, *246*
Weinman, M. L., 95, *234*
Weisel, T. N., 217
Wells, G. A., 205, 209, *246*
Wernicke, C. (Wernicke's area), 19, 25, 49–50, 98, 114, 161, *246*
Wexler, B. E., 19, 21, 26, *246*
Wexler, K., 115, 240
Wheelwright, J. B., 131, *234*
Widen, L., 51, *233*
Wiener, M. S., 139–140, *230*
Will, O., 107
Wilson, D. H., 24, *237*
Winnicott, D. W., 107
Winshaw, I. Q., 22, 87, 113, *236*
Winson, J., 56–57, *246*
Wohlman, M., 138, *230*
Wojcik, M., *240*
Wolff, P. H., 217, *246*
Wright, C. V. E., 109, *229*

Y

Yakovlev, P. I., 188, *247*
Young, J. Z., 47, 49, 209, *247*

Z

Zaide, J., 44, 239
Zaidel, D., 28, *247*
Zec, R. F., 95, *246*
Zetzel, E., 7, *247*
Zook, J. M., *239*

For Product Safety Concerns and Information please contact our EU representative GPSR@taylorandfrancis.com Taylor & Francis Verlag GmbH, Kaufingerstraße 24, 80331 München, Germany